This series fosters information exchange and discussion on management and industrial engineering and related aspects, namely global management, organizational development and change, strategic management, lean production, performance management, production management, quality engineering, maintenance management, productivity improvement, materials management, human resource management, workforce behavior, innovation and change, technological and organizational flexibility, self-directed work teams, knowledge management, organizational learning, learning organizations, entrepreneurship, sustainable management, etc. The series provides discussion and the exchange of information on principles, strategies, models, techniques, methodologies and applications of management and industrial engineering in the field of the different types of organizational activities. It aims to communicate the latest developments and thinking in what concerns the latest research activity relating to new organizational challenges and changes world-wide. Contributions to this book series are welcome on all subjects related with management and industrial engineering. To submit a proposal or request further information, please contact Professor J. Paulo Davim, Book Series Editor, pdavim@ua.pt

More information about this series at http://www.springer.com/series/11690

Kaushik Kumar · Divya Zindani ·
J. Paulo Davim

Mastering SolidWorks

Practical Examples

 Springer

Kaushik Kumar (ORCID)
Department of Mechanical Engineering
Birla Institute of Technology
Jharkhand, India

Divya Zindani (ORCID)
Married Scholars Hostel
National Institutes of Technology
Silchar, Assam, India

J. Paulo Davim (ORCID)
Department of Mechanical Engineering
University of Aveiro
Aveiro, Portugal

ISSN 2365-0532   ISSN 2365-0540 (electronic)
Management and Industrial Engineering
ISBN 978-3-030-38900-0   ISBN 978-3-030-38901-7 (eBook)
https://doi.org/10.1007/978-3-030-38901-7

This Springer imprint is published by the registered company Springer Nature Switzerland AG
The registered company address is: Gewerbestrasse 11, 6330 Cham, Switzerland

Preface

Engineering Graphics, more commonly known worldwide as *Language of Engineers*, is the pictorial representation of an object usually prepared using drafters (or mini drafters), pencil, T-square, set squares, and scales, and the permanent versions and tracings were made by ink. Physical objects are of three-dimensional nature but are required to be represented on a 2D drawing sheet. This led to the development of various views (full, half sectional, and full sectional) for complete picture. Hence, it became a subject of expertise and to 'read' an engineering drawing was not a common man's cup of tea. Organizations had to recruit and retain expert draftsmen for the translation of an engineering drawing for the manufacturing and production units.

With the invention of computers and the development of different software, engineering practice also changed. As the manual drawings required large workspace and hardware, consumed more time and labor, need strenuous effort in editing or modifying drawings, and pose a lot of storage or transfer problems, the computer-based graphic system came as a solution to produce fast, simple, accurate, and repeatable engineering drawings. The Computer-Aided Drafting (CAD) software was developed under two categories: coordinate-based system and parametric-based system. Coordinate-based software (AutoCAD, STADD, etc.,) created the object with x-, y- and z-coordinates, making the object securely placed at a specific point with respect to the origin whereas parametric-based software (SOLIDWORKS, CATIA, etc.,) used dimensional parameters, like length, radius, etc., for object creation. For 2D modeling, coordinate-based software is preferred, and hence, architectural drawing extensively uses them. But while working with 3D modeling coordinate-based software becomes quite difficult to handle specially when editing is required. The parametric-based software is preferred as they do not specify coordinate; hence, models get updated just by changing the dimensions. This feature has increased their acceptability to engineering and industrial community. Further to modeling ease, features like Computer-Aided Manufacturing (CAM) and Rapid Prototyping (RP) also require output from parametric-based software.

Due to the increasing acceptability, today the software market is flooded with 3D modeling software, namely SOLIDWORKS, CATIA, CREO, UNIGRAPHICS, MECHANICAL INVENTOR, etc. Although the basic features offered by all of these are more or less same, SOLIDWORKS has become the most popular due to its ease of operation and more user friendly. Hence, most of the industries and academic institutions, both globally and locally, prefer the same as their working tool.

It is true to say that in many instances, the best way to learn complex behavior is by means of imitation. For instance, most of us learned to walk, talk, run, etc., solely by imitating the actions and speech of those around us. To a considerable extent, the same approach can be adopted to learn using SOLIDWORKS software by imitation, i.e., using the examples provided in this book. This is the essence of the philosophy and the innovative approach used in this book. The authors have attempted in this book to provide a reader with a comprehensive cross section of various models in a variety of engineering areas, in order to provide a broad choice of examples to be imitated in one's own work. In developing these examples, the authors' intent has been to utilize most of the features and templates provided by the developer. By displaying these features in an assortment of disciplines and modeling, the authors hope to give readers the confidence to employ these program enhancements in their own applications. The primary aim of this book is to assist in learning the use of SOLIDWORKS software through examples taken from various areas of engineering. The content and treatment of the subject matter are considered to be most appropriate for university students studying engineering and practicing engineers who wish to learn the use of SOLIDWORKS. This book is exclusively structured around SOLIDWORKS, and no other solid modeling software currently available is considered.

This book contains 14 chapters in which 12 chapters describe the various features of the software whereas the last two is dedicated in drafting two complete engineering components which are quite complicated but very popular among students, researcher, and industrialists.

Chapter 1 introduces the readers to the requirement of 3D modeling system and then walks into the arena of SOLIDWORKS. The chapter also highlights the necessity of the same in solving engineering problems.

The basic feature for creating a solid is the 2D sketch, which acts as the base for the development of the solid. Chapter 2 elaborates on the same.

Chapter 3 deals with basic sketch relationships and offers the reader with features like dimensioning.

Chapter 4 concentrates on part modeling or solid modeling. The various features available are introduced, and by the help of screenshots, the user is educated regarding usage of the features.

In the next three chapters, i.e. Chapters 5–7, the reader moves further to explores the various features associated with model development, editing, etc. Here, thin features are explained in Chap. 7 with the help of screenshots.

Chapters 8 and 9 introduce sheet metalworking. Apart from explaining regular features associated with sheet metal forming, there are advanced features like the addition of two bends, curling of the edges, attachment of two edges with the help of gusset and also of creating gap between the edges, etc. All the above said can be done in sheet metal which is discussed one by one in Chap. 9.

Next two chapters, i.e. Chapters 10 and 11, explain assembly of components. Chapter 10 provides exposure to assembly of round parts by aligning the axes of rotation whereas Chap. 11 works with rectangular components aligning x-, y- and z-axes.

Chapter 12 concentrates on the development of 2D drawings from 3D modeled components.

As stated earlier Chap. 13 and 14 provide drafting two complete engineering components from creating the components to the assembly. In Chap. 13, flanged coupling is considered whereas in Chap. 14 Footstep bearing is taken up. The two examples are chosen as they are quite popular among the undergraduate students doing the courses of engineering graphics, machine drawing, and CAD.

There is a famous saying 'Seeing is Believing', so from exclusive treatment to all aspects of SOLIDWORKS, would help the reader to navigate various tools and utilities of SOLIDWORKS used. The strategy used in this book, i.e., to enable readers to learn SOLIDWORKS by means of imitation, is quite unique and very much different than that in other available books on SOLIDWORKS. The strategy used in this book, i.e., to enable readers to learn SOLIDWORKS by means of imitation, is very much different than that in other available books on SOLIDWORKS.

The authors are of the opinion that the planned book is very timely as there is a considerable demand, primarily from university engineering course, for a book which could be used to teach, in a practical way, SOLIDWORKS—a premiere solid modeling software. Also practicing engineers are increasingly using SOLIDWORKS for product development and product realization using CAM or RP Techniques, hence the need for a book which they could use in a self-learning mode.

First and foremost, we would like to thank God. In the process of putting this book together, it was realized how true this gift of writing is for anyone. You have given the power to believe in passion, hard work, and pursue dreams. This could never have been done without the faith in You, the Almighty. We would like to thank all of our colleagues and friends in different parts of the world for sharing of ideas in shaping our thoughts. We are grateful to all design managers whose kind contribution helped in shaping this. Our efforts will come to a level of satisfaction if the professionals concerned with all the fields related to design processes will get benefitted. We owe a huge thanks to all of our technical reviewers, editorial advisory board members, book development editor, and the team of Publisher Springer Nature for their availability for work on this huge project. All of their efforts helped to make this book complete, and we could not have done it without them.

Throughout the process of writing this book, many individuals, from different walks of life, have taken time out to help us out. Last, but definitely not least, we would like to thank them all, our well-wishers, for providing us encouragement. We would have probably given up without their support.

All Images and models developed in this text were created using SOLIDWORKS. SOLIDWORKS is a registered trademark of Dassault Systèmes.

Jharkhand, India   Kaushik Kumar
Silchar, India   Divya Zindani
Aveiro, Portugal   J. Paulo Davim

Contents

About the Authors

Kaushik Kumar B.Tech (Mechanical Engineering, REC (Now NIT), Warangal), M.B.A. (Marketing, IGNOU), and Ph.D. (Engineering, Jadavpur University), is presently Associate Professor in the Department of Mechanical Engineering, Birla Institute of Technology, Mesra, Ranchi, India. He has 14 years of Teaching and Research and over 11 years of industrial experience in a manufacturing unit of global repute. His areas of teaching and research interest are quality management systems, optimization, non-conventional machining, CAD/CAM, rapid prototyping, and composites. He has nine patents, 28 books, 19 edited book volume, 43 book chapters, 137 international journal, 21 international, and one national conference publications to his credit. He is on the editorial board and review panel of seven international and one national journals of repute. He has been felicitated with many awards and honors.

Divya Zindani B.E. (Mechanical Engineering, Rajasthan Technical University, Kota), M.E. (Design of Mechanical Equipment, BIT Mesra), presently pursuing Ph.D. in Department of Mechanical Engineering, National Institute of Technology, Silchar. He has over 2 years of industrial experience. His areas of interest are optimization, product and process design, CAD/CAM/CAE, and rapid prototyping. He has one patent, five books, five edited book, 20 book chapters, and 11 international journal publications to his credit. He has been felicitated with awards.

J. Paulo Davim received his Ph.D. degree in Mechanical Engineering in 1997, M.Sc. degree in Mechanical Engineering (materials and manufacturing processes) in 1991, Mechanical Engineering degree (5 years) in 1986, from the University of Porto (FEUP), the Aggregate title (Full Habilitation) from the University of Coimbra in 2005, and the D.Sc. from London Metropolitan University in 2013. He is Senior Chartered Engineer by the Portuguese Institution of Engineers with an M.B.A. and Specialist title in Engineering and Industrial Management. He is also EUR ING by FEANI-Brussels and Fellow (FIET) by IET-London. Currently, he is Professor at the Department of Mechanical Engineering of the University of Aveiro, Portugal. He has more than 30 years of teaching and research experience in manufacturing, materials, mechanical and industrial engineering, with special emphasis in machining and tribology. He has also interest in management, engineering education, and higher education for sustainability. He has guided large numbers of postdoc, Ph.D., and master's students as well as has coordinated and participated in several financed research projects. He has received several scientific awards. He has worked as evaluator of projects for ERC-European Research Council and other international research agencies as well as examiner of Ph.D. thesis for many universities in different countries. He is Editor-in-Chief of several international journals, Guest Editor of Journals, Books Editor, Book Series Editor, and Scientific Advisory for many international journals and conferences. Presently, he is an Editorial Board Member of 30 international journals and acts as reviewer for more than 100 prestigious Web of Science Journals. In addition, he has also published as editor (and co-editor) more than 125 books and as author (and co-author) more than 10 books, 80 book chapters, and 400 articles in journals and conferences (more than 250 articles in journals indexed in Web of Science core collection/h-index 53+/9000+ citations, SCOPUS/h-index 57+/11,000+ citations, Google Scholar/h-index 75+/18,000+).

Chapter 1
Introduction

1.1 Graphic Language

An individual can utilize graphic language as effective method for communications with others for passing on ideas and information on specialized issues. However, for compelling trade of ideas and information with others, the engineer must have capability in (i) dialect, both composed and oral, (ii) images and symbols related with fundamental sciences, and (iii) the graphic language. Engineering drawing is a reasonable graphic language from which any trained individual can visualize the required object. As engineering drawing shows the correct picture of an object, it clearly passes on similar ideas to each trained eye. Regardless of dialect, the drawings can be adequately utilized in different nations, in addition to the nation where they are made. In this way, the engineering drawing is the universal language and dialect of all engineers around the globe. Engineering drawing has its source at some point in 500 BC in the administration of King/Pharaohs of Egypt when images were utilized to pass on the ideas among individuals.

1.2 Significance

The graphic language had its existence when it became important to assemble new structures and make new machines or something like that, notwithstanding speaking to the current ones. Without graphic language, the ideas and information on specialized issues must be passed on by writing or through speech; both are

Electronic supplementary material The online version of this chapter (https://doi.org/10.1007/978-3-030-38901-7_1) contains supplementary material, which is available to authorized users.

© Springer Nature Switzerland AG 2020
K. Kumar et al., *Mastering SolidWorks*, Management and Industrial Engineering, https://doi.org/10.1007/978-3-030-38901-7_1

problematic and hard to comprehend by the shop floor individuals for assembling. This technique includes hefty amount of time and work, as well as assembling errors and blunders. Without engineering drawing, it would have been difficult to create objects, for example, airships, vehicles, trains, and so on, each requiring a large number of various parts.

1.3   Introduction to Drawings

The pictorial representation of an object is one of the simplest and common communication modes adopted by the mankind. When the need arose to develop and manufacture a product for human use, the formal pictorial representations of the products got refined as engineering drawing and became a common communication tool among the engineers. The guidelines for engineering drawings were standardized universally, and hence, the international and national standards were formulated. The drawings can be made traditionally either by manual means or by the digital means due to the recent advancement of computers. In order to practice and understand the drawing effectively, this chapter discusses the preparation of engineering drawings with manual drafting tools with the incorporation of the dimensioning and lettering formalities as recommended in Bureau of Indian Standards (BIS).

1.4   Modes of Communication

Humans tried to develop different modes of communication with each other since they started to exist. Early humans tried to live in groups, in order to protect them from attack. To communicate among their groups, the early people needed a mode of communication. Initially, sound was used as the communication media. The sound produced by one group was understood 'by the other and was replied again in similar manner'. However, during nights, the light source in the form of fire became effective to convey the presence of one group to the other. When they moved closer to each other and when their civilization improved, communication through signs started evolving. The practice and development of sign conventions with sound components added later, turned out to be the language, and became vernacular depending on the groups that practiced it. Apart from these modes, the mankind had started communicating with the sketches or pictures of the animals, the nature, landscape, and the environment in which they lived and used the available tools, such as charcoal. The walls of public places or tombs or places of worship were used for this purpose. Understanding the sizes and shapes of the objects and incorporating them into the drawings with due proportions has led to the development of paintings and sculptures. The artists further improvised them to convey the feel or the emotion. This had received wide popularity among the mankind. Though vernacular languages were developed later for effective communication, they got restricted to that group of people only. Hence, the pictorial representations were the only common mode of communication because of their uniqueness, simplicity, and reach to the entire mankind, universally, irrespective of the differences in the topography, language, and knowledge level of the people. The power of pictorial communication had been understood by the engineers and was further improvised to reveal the shape and sizes of the engineering objects to be created. The three-dimensional objects are presented in two-dimensional drawings, which are known as engineering drawings. Since the main purpose of engineering drawing is to create or manufacture the object for human use; these drawings are to be realistic with the true dimensions, unlike the other pictorial drawings created by the artist, whose focus is only toward the appreciation and not the realism. The engineering drawings can be further tuned to make the shapes aesthetically pleasing and ergonomically viable. Since at the universal level, the facts conveyed by means of such drawings are to be the same, and the drawings are drawn using the universal standardization procedures and specifications.

1.5   Engineering Drawing as a Communication Tool

The history of drawing is as old as the history of mankind. People started sketching or drawing much before they learnt how to write, as that helped them to communicate. The oldest known cave art paintings are nearly 35,000 years old. In the prehistoric age, charcoal and ochre were used to create the images of animals. Ancient Egyptians (about 3000 BC) decorated the walls of their temples and tombs with the drawings of Gods. Greeks during 800 BC recorded their drawings on pottery surfaces and later, after 500 BC, made them more realistic with natural proportions and details, and such figures were known as silhouettes. Artists made drawings on prepared animal skins such as vellum or parchment and on tablets of wood, wax, or slate, before paper was invented. Sometimes, the drawings were also made directly on the walls to be painted. Roman artists painted scenes from the stories and the 'great scrolls' on the walls of their homes. For centuries, engineering was focusing on war, either building defensive fortifications or the machines to attack these fortifications. The first non-military engineering discipline came to exist was known as the 'civil engineering.' The sketches of war machines that existed during that period were made on parchment or scratched on clay tablets. However, there is no record of sketches that existed for the magnificent construction of the Colosseum in Rome or the Greek Pantheon. In the early fifteenth century, the concept of graphic projections was well understood by Italian architects and the drawing took a rebirth, and at the same period, the invention of paper began to replace the parchment as a drawing medium. During the Italian Renaissance (1400–1525), Flemish Renaissance (1440–1540), German Renaissance (1400–1540), and High Renaissance (1500–1520) periods, drawing became the foundation work in all arts. Art students were trained first in drawing, before taking up painting, sculpture, and architectural works. Portrayal of human figures became more realistic due to this. Adoption of the prominent Golden Rectangle or Golden Mean Ratio (3:5) by Leonardo da Vinci, the mathematician gave more realistic drawings. Most of the drawings drawn in these periods are housed in European museums and libraries and are restricted to academic researchers. These drawings are mostly in the form of sketches, are not to scale, and do not have dimensions but contain exhaustive textual descriptions. Some of the best-known engineering drawings are that of Leonardo da Vinci. Though he is well known for the Mona Lisa painting, he was a designer of military machines, a mathematician, an artist, and a scientist. His design works were artistic in nature and did not have multiple views, but still craftsmen were able to construct models from his single-view sketches. Simultaneously, people who practiced drawing realized the need to make it stand on its own merits and wanted to incorporate greater precision. In 1435–1436, Leon Battista Alberti wrote two books and explored the need to incorporate mathematics as a common ground into contemporary drawings. Proposals for drawings with multiple views emerged. With the works of Rene Descartes (1596–1650) and Gaspard Monge (1746–1818) on the development of descriptive geometry, engineering drawing began to evolve in the eighteenth century and picked up speed with the industrial revolution of the nineteenth century. Peter Booker in his book A History of Engineering Drawing distinguished the industrial/technical practices and the earlier craft practices. The growth of the patent process really became the catalyst to develop technical drawing. When the need arose for the submission of multiple copies of drawings for obtaining patent, the invention of the blueprinting device (cyanotype process) took place in 1842 by Sir John Herschel. Industrial drawings were prepared using pencil, T-square, set squares, and scales; permanent drawings and tracings were made by ink. Then, the attention of lettering was realized and lettering templates came into existence. After World War I, the first American Drafting Standard was prepared in 1935, with this focus. A major advance took place on the drafting device that led to the Universal Drafting Machine. This device was basically developed to combine the 'T-square, set squares, scales, and protractors, which had allowed different engineering disciplines to develop their own approaches to design and drafting'. While architects followed an appreciable style of drafting for their works, aeronautical engineers used a different style with the purpose of accuracy. Not only the drafting quality, but also the speed with which the drawings were produced, also became important. After World War II, together with a new generation of reproduction machines, the time taken to prepare drawings was substantially reduced.

1.6 Need for Drawings

The drawings arranged by any specialized individual must be clear, unmistakable in importance and there ought not be any degree for in excess of one translation, or else litigation may emerge. In various dealings with contracts, the illustration is an official report and the achievement or disappointment of a structure relies upon the lucidity of points of interest given on the illustration. Along these lines, the drawings ought not give any degree for error even coincidentally. It would not have been conceivable to deliver the machines/vehicles on a mass scale where various assemblies and sub-assemblies are included, without clear, right, and exact drawings of the same. To accomplish this, the specialized individual must pick up an intensive information of both the standards and customary routine with regard to draughting. On the off chance that these are not accomplished and additionally rehearsed, the drawings prepared by one may pass on various significant meanings to others, causing pointless postponements and costs various expenses in production shops. Henceforth, a designer ought to have great information, in setting up a right drawing as well as to peruse the drawing effectively.

1.7   Drawing

The graphical portrayal of any object or thought can be named as drawing. A drawing can be prepared either utilizing free hand or utilizing designing instruments or utilizing PC program.

1.8   Types of Drawing

1. Artistic Drawing
2. Engineering Drawing.

1.8.1   Artistic Drawing

The illustration speaking to any object or thought which is outlined in free hand utilizing imagination of craftsman and in which appropriate scaling and dimensioning is not kept up is called an artistic drawing. Case: painting, posters, expressions, and so forth.

1.8.2   Engineering Drawing

Engineering drawing can be characterized as a graphical dialect utilized by engineers and other specialized individuals related with the engineering profession which completely and plainly characterizes the necessities for built things. It is a two-dimensional portrayal of a three-dimensional question. We can say that the art of representing a genuine or nonexistent object accurately utilizing a few designs, images, letters, and numbers with the assistance of engineering drawing instruments is called engineering drawing. The craft of speaking to designing objects, for example, structures, streets, machines, circuits, and so forth on a paper is called engineering drawing. It is utilized by architects and technologists. An engineering drawing gives all data about size, shape, surface, materials, and so forth of the object. Case: building drawing for structural designers, machine drawing for mechanical specialists, circuit charts for electrical, and electronics engineers.

Table difference between artistic and engineering drawing

Artistic Drawing	Engineering Drawing
Purpose of artistic drawing is to convey emotion or artistic sensitivity in some way	Purpose of engineering drawing is to convey information about engineering object or idea
Can be understood by all	Need some specific knowledge or training to understand
Scale maintaining is not necessary	Scale maintaining is necessary
No special requirement of engineering instruments	Engineering drawing instruments are used to make the drawing precise
An artistic drawing may not be numerically specific and informative	An engineering drawing must be numerically specific and informative
Standard drawing code need not to be followed	Standard drawing code (like ISO, ANSI, JIS, BS, etc.,) must be maintained

1.9   Purpose of Engineering Drawing

It is extremely troublesome and complex to clarify some specific designing necessities in word. In such cases, all around dimensioned and appropriately scaled drawings can make it straightforward that for specialized workforce. Building drawing fills this need. Any item that will be produced, created, amassed, developed, constructed, or subjected to some other kinds of change process should first be planned. To make the result from the plan justifiable to any third party, engineering drawing is the most ideal way.

1.10   Applications of Engineering Drawing

Engineering drawing is a fundamental part of all building ventures. Some imperative employments of engineering drawing are specified underneath:

1. It is utilized as a part of ships for route.
2. For assembling of machines, vehicles, and so forth.
3. For the development of structures, streets, bridges, dams, electrical and media transmission structures, and so forth.
4. For assembling of electric machines like TV, telephone, PCs, and so on.

1.11 Types of Engineering Drawing

Engineering drawing can be grouped into the following four major categories:

1. Geometrical Drawing

   a. Plane geometrical drawing
   b. Solid geometrical drawing

2. Mechanical Engineering Drawing
3. Civil Engineering Drawing
4. Electrical and Electronics Engineering Drawing, etc.

1.12 Classifications

1.12.1 Machine Drawing

It is relating to machine parts or segments. It is exhibited through various ortho-graphic perspectives, with the goal that the size and shape of the segment are completely comprehended. Part drawings and assembly drawings have a place in this grouping/classification.

1.12.2 Production Drawing

A production drawing, likewise alluded to as working drawing, ought to have all dimensions and finishing procedures, for example, heat treatment, sharpening, lapping, surface finish, and so forth, to direct the specialist on the shop floor in delivering the component. The title ought to likewise say the material utilized for the item, number of parts required for the assembled unit, and so on.

Since a skilled worker will customarily make one part at any given moment, it is advisable to set up the production drawing of every component on a different sheet. However, in many cases the drawings of related parts might be given on a similar sheet.

1.12.3 Part Drawing

Segment or part drawing is a detailed drawing of a segment to encourage its fabrication or manufacturing. Every one of the standards of orthographic projection and the strategy of graphical portrayal must be taken after to impart the points of interest in a part drawing. A part drawing with production details of elements is appropriately called as a production drawing or working drawing.

1.12.4   Assembly Drawing

A drawing that demonstrates the different parts of a machine in their right working areas is an assembly drawing. There are a few sorts of such drawings.

1.13   Emergence of Engineering Graphics

When computers came into existence, they changed the engineering practice too. As the manual drawings require large workspace and hardware, consume more time and labor, need strenuous effort to edit the drawings, and pose a lot of storage or transfer problems, the computer-based graphic system was realized to produce fast, simple, accurate, and repeatable engineering drawings. The computer-aided drafting (CAD) procedures facilitate all these benefits. Due to the rapid growth in mathematical procedures and computing algorithms, three-dimensional models are drawn with ease. Using primitive shape of basic solids and Boolean combinations, three-dimensional solid models are created. The software 'solid modeling' is used to achieve these drafting. From the solid modeling, the required two-dimensional drawings of objects for any specified position can be obtained. In addition, the application of graphical techniques helps the solid models to get rotated or turned to any position, amenable for the addition and deletion of various parts and obtaining the corresponding two-dimensional drawings. The method of preparing and altering drawings through computers using graphical techniques including animation effects is known as 'Engineering Graphics.'

1.14   Computer-Aided Drawing and Drafting

The most downside with conventional drawing is data sharing, i.e., on the off chance that an engineer is drawing outline of machine segment and all of a sudden the maker needs to alter measurement of deepest piece of the segment; in such circumstances, one cannot adjust the drawing effectively drawn, he ought to redraw the segment. CADD is an electronic tool that empowers us to make brisk and precise drawings with the utilization of a PC. Drawings made with CADD have various advantages over drawings made on a drawing board. CADD drawings are perfect, clean, and exceptionally presentable. Electronic drawings can be adjusted effortlessly and can be displayed in a variety of forms. There are several CADD programs accessible in the CADD business today. Some are expected for general drawing work while others are centered around particular building applications. There are programs that empower you to complete 2D drawings, 3D drawings, renderings, shadings, building figuring, space arranging, auxiliary outline, piping formats, plant configuration, project management, and so on.

Examples of CAD software are AutoCAD, PRO-Engineer, IDEAS, UNIGRAPHICS, CATIA, SOLIDWORKS, etc.

1.15   History of CAD

In 1883, Charles Babbage created an idea for personal computing. To begin with CAD was first given by Ivan Sutherland (1963). After a year, IBM delivered the main CAD framework. Numerous progressions have occurred from that point forward, and with more powerful PCs, it is presently conceivable to do every one of the designs utilizing CAD including two-dimensional drawings, strong displaying, complex designing, production, and assembling. New innovations are continually imagined which make this procedure speedier, more adaptable, and all the more powerful.

1.16   CAD Background

In a little over a generation's time, the methods used to create technical drawings have fundamentally changed from using pencil and paper to the use of computer-aided drafting, better known as CAD. The analog world of drafting boards, T-squares, triangles, and even the romantic French curve has given way to the brave new digital world of computers. No longer must you refill your mechanical pencil when you run out of lead, find your eraser when you make a mistake, or walk across the room to share a design with another person. Using CAD, you can draw something once and copy it hundreds, or even thousands, of times. Changing a design can be as simple as pushing a button. Drawings can be shared instantaneously across the room or even around the world over a computer network.

1.17   Advantages of CAD

These and the other benefits of CAD include the following:

- Detail drawings might be made all the more rapidly, and rolling out improvements is more effective than remedying drawings drawn physically
- It permits diverse perspectives of a similar object and 3D pictorial view, which gives better representation of drawings

- Designs and images can be put away for simple review and reuse
- By utilizing the PC, the drawing can be created with more precision
- Drawings can be all the more helpfully documented, recovered, and transmitted on plates and tape
- Quick Design Analysis, likewise Simulation and Testing Possible
- Increased productivity

  - Drawing content can be continuously reused
  - Text and dimensions can be created and updated automatically
  - Hatch and pattern fills can be placed with a single pick
  - Revising and editing drawings can be done quickly with minimum effort

- Improved precision

  - Digital information is accurate to 14 decimal places
  - Geometry is precisely located using the Cartesian coordinate system
  - It is possible to snap to control points and features on existing drawing geometry to accurately locate drawing information
  - Polar and object tracking features can be utilized for precise angular measurements

- Better collaboration

  - Drawings can be shared across a network (locally and globally)
  - Drawings can be referenced and updated in real time with notification
  - Revisions and markups can be managed electronically via email and Internet-based document management systems

- 3D visualization and analysis
- 3D animations and walk-throughs can be easily generated to allow you and potential clients to visualize a design before it is constructed
- Interference checking can be done to ensure that parts do not run into each other before they are created
- Engineering calculations such as finite element analysis (FEA) and other structural calculations can be performed automatically
- Computer prototypes can be created and tested, eliminating the time and materials needed to manufacture a real-world prototype
- Reduced storage room
- Corrections can be made effortlessly
- Repetitive parts of the drawing can be spared and foreign made as a major aspect of a 'computer-aided design library'
- CAD frameworks can be connected with CAM machines to deliver questions straight from the drawings
- 3D CAD plans can be made to look reasonable by utilizing the material library for customers to see
- CAD outlines can be effectively shared between organizations and division utilizing email
- CAD can be utilized to make reproduced situations to demonstrate the customer.

1.18   CAD Disadvantages

- Work can be lost if the computer crashes
- Work could be defiled by viruses
- Work could be stolen or 'hacked'
- Time taken to figure out how to utilize the product
- Initial expenses of purchasing a high-performance computer are high
- Time and cost of training staff
- Continual requirement for refreshing programming or working frameworks
- CAD/CAM systems mean less individuals should be employed.

1.18.1   AutoCAD

AutoCAD is a commercialized computer-aided design (CAD) and drafting software application. Created and promoted via Autodesk, AutoCAD was first introduced in December 1982 as a work area application running on microcomputers with interior drawings controllers. Before AutoCAD was presented, most business CAD programs kept running on mainframe computer PCs or minicomputers, with every CAD administrator (client) working at a different design terminal. Since 2010, AutoCAD was introduced as a mobile platform and web application also, showcased as AutoCAD 360. AutoCAD is utilized over an extensive variety of enterprises, by modelers, project supervisors, engineers, graphic designers, and numerous different experts. It was supported by 750 instructional hubs worldwide in 1994.

1.18.2   Advantages of Using AutoCAD

- A mid-go software
- User friendly
- Fulfills the vast majority of the expert needs of a mechanical engineer
- Lots of time savings (yet not generally)
- Environment friendly
- Ease of accuracy—e.g., joining of lines straight on is not conceivable physically with wanted precision
- Easy to alter
- Ease of repetition/addition
- Reproduction of drawing is simple, quick, and solid as far as dimensional exactness and accuracy is concerned
- 2D and 3D generations of drawings are simple

- Zoom command facility enables us to see the minute details of elements on bigger scale
- A facility of modeling and making parts library is accessible
- Parent–child relationship permits different altering in single order and implies in the event that when one section is altered, all other connected parts are auto altered
- Data taking care of information stockpiling and information arranging turn out to be simple in delicate frame when contrasted with the traditional printing and documenting
- Data transmission to far off and remote regions is simple and conservative through messaging and connection sharing
- Layers alternative enable us to stow away or demonstrate some particular subtle elements of a mind-boggling get together drawing for clear understanding
- There is no restriction of a drawing size. You can draw a two light-years long queue through legitimate scale stetting
- Auto scale choice is accessible. Full-scale drawings can be shaped
- Geometrical relationship can make brisk scientific tasks simple; e.g., you can make a recorded circle utilizing measurements of the triangle or polygon and so forth
- BOM or BOQ can be delivered with precision
- Mass, area, volume, center of gravity can be ascertained
- Auto dimensioning—simple, exact, and quickly done inside no time
- Dimension standards are entirely taken after
- Bullion task is conceivable
- Copy, scale, move, stretch, turn, pattern orders make the errand straightforward and simple
- Draw guiding through O-snap, Ortho, and so on
- Direct and quick drawing of symmetrical articles, e.g., circle, oval, polygon, rectangles, triangles, and so on
- Geometrical relationships are conceivable, like two parallel lines, tangent lines, opposite lines, and so forth
- Fixed line thickness all through the drawing—recollect that, 'of course, the line thickness of auto Cad is zero'
- Fill, hatch, section lines, chamfer, and filet orders make the product a gift as these tasks are exceptionally troublesome in manual drafting
- Adding text—lettering—is a major facility
- Images can be imported and allowed for the digitization of regular outlines and drawings.

1.18.3   Disadvantages of Using AutoCAD

- Expensive hardware is required like computer, plotter, extra-large screen, and so forth
- Registered programming is costly, and it requires a substantial re-happening yearly expense
- Equipment is delicate and can be harmed definitely
- Continuous updating of the hardware and programming is required
- Electricity is compulsory for running the hardware which is a major issue for underdeveloped nations like SAARC countries which are confronting a serious power emergency nowadays
- Data stockpiling is additionally delicate—high likelihood of information defilement; information misfortune is a major risk for all delicate soft form information banks. To conquer this potential fiasco, specialists propose 'Mirroring of Data' that is capacity of information on multiple hard drives on various areas which include a major capital allotment
- Piracy and hacking dangers are dependably there when you utilize web for information exchange or even capacity of information on a PC which is associated with web
- Special PC Skills are required
- Proper maintenance, supervision, and organization are required for PC organizing.

1.19   Limitations of AutoCAD

Autodesk AutoCAD is a standout among the most mainstream computer-aided design (CAD) projects, and it makes precise, proficient drawings. Nonetheless, the program misses the mark for computer modeling and graphic designing. The application has modeling tools and also shading and fill tools, yet AutoCAD does not contrast well with contemporary building information modeling (BIM), three-dimensional modeling software.

1.19.1   Line

AutoCAD produces drawings utilizing line and shape instruments. Curves, circular segments, and straight lines deliver the shapes; however, AutoCAD cannot alter the line and area as unreservedly as illustration programs—altering and covering lines and line weights are constrained to a couple of alternatives. What is more, AutoCAD makes drawings from just lines, never volumetric models, for example, with BIM. In any case, the application can deliver exact three-dimensional geometry with constrained material impacts.

1.19.2 Constrained File Formats

Since AutoCAD is one of the main CAD programs, it restrains the quantity of document groups it can import or export, in light of the fact that Autodesk anticipates that different projects will fare to AutoCAD positions, for example, DWG and DXF. Tragically, this makes problems when utilizing different projects with all the more intense tools and sending out the program to an AutoCAD format—geometry, shading, and impacts are lost frequently.

1.19.3 Shading, Fill, and Texture

AutoCAD drawings and models can have shading, fill, and surface, utilizing the line and hatch tools. Nonetheless, as far as possible the quantity of conceivable colors to 256 and the hatching gives just a modest bunch of surfaces, so you cannot make photograph practical pictures like representation programs. Rather, you can import image files and can make material maps for AutoCAD renderings, yet AutoCAD's rendering capacities cannot contend with three-dimensional modeling projects or representation programs.

1.19.4 Non-parametric

AutoCAD gives apparatuses to make three-dimensional models; however, altering the models requires numerous steps, not at all like BIM parametric models, which naturally change the greater part of the model segments while altering components. Besides, data is not connected to the models, for example, with BIM parametric models—BIM gives the designer information about the material and volumetric properties of the building project.

1.20 SOLIDWORKS

SOLIDWORKS is a solid model computer-aided design (CAD) and computer-aided engineering (CAE) software program that runs on Microsoft windows. SOLIDWORKS is distributed by Dassault Systèmes. As per the distributer, more than two million architects and designers in excess of 165,000 organizations were utilizing SOLIDWORKS.

Similarly, as the move from 2D design tools to 3D CAD solutions product improvement, the development of 3D model-based definition (MBD) incorporated assembling innovation offers generous efficiency points of interest over the utilization of conventional 2D engineering drawings to drive manufacturing. MBD drawing less manufacturing solutions expand the advantages of 3D configurations to assembling and manufacturing, including time and cost savings—through enhanced, all the more firmly coordinated communication of product and manufacturing information (PMI) for production—and also decreased scrap/revamp, enhanced precision, and quicker throughput. SOLIDWORKS MBD programming computerizes the generation, association, customization, and sharing of PMI information required for generation in simpler to see, more data-rich 3D digital formats, for example, SOLIDWORKS parts and assemblies, eDrawings, and 3D PDF, giving item improvement and assembling activities a huge focused edge.

1.20.1 Advantages of SOLIDWORKS

- Reduces acquirement cost and accomplishes faster supplier reaction through model-based correspondence
- Cuts tooling and creation costs
- Removes the time required to tidy up 2D drawings
- Reduces time required to make and keep up assembling documentation
- Reduces the quantity of comments and measurements
- Reduces designing changes
- Streamlines the ECO procedure
- Reduces the quantity of files for close down
- Reduces errors
- Removes model to drawing error
- Manipulates the 3D show straightforwardly for clearness
- Cuts scrap and adjust
- Complies with rising industry principles
- Fits into an organization's as well as supply chains current procedures consistently
- Exports broadly acknowledged industry positions, including 3D PDF, eDrawings, and SOLIDWORKS records
- SOLIDWORKS peruses DWG files into representations or 2D drawings with the goal that clients can manufacture 3D models in light of inheritance information
- Saves time squandered on re-displaying parts in light of 2D drawings
- Shares and archives wise and natural 3D information specifically to models, PMI, perspectives, and meta properties with the goal that clients and their suppliers do not need to sit around idly in re-modeling parts in light of customary 2D drawings. Overhauls existing items to new models substantially faster and plans tooling and fixtures all the more effortlessly in view of 3D information.

1.20.2 Introducing SOLIDWORKS

From this point onward, we will be dealing with the key features of the parametric solid software, difference between the sketched and applied features and convey about the difference dimensioning and design intent of the software.

The main aim is to know how to use mechanical design software, create different part assemblies, and then to make simple drawings with the help of those parts and assemblies. SOLIDWORKS is sturdy and firm in applications that are visionary in covering each and every minute detail and distinct feature of the software and still have the course to be an extreme length. So, the focus is only on the fundamental skills and concepts serving as an essential component to the successful use of SOLIDWORKS.

SOLIDWORKS mechanical design act of implementing the control of equipment with advanced technology software is a *feature-based parametric solid modeling* design tool which makes advantage of the easy to gain knowledge related to windows^{TM} graphical user interface. One can produce fully characterized 3D solid models with or without the presence of constraints while employing minimal human intervention or user-defined relations to succeed in representing design intent.

There are few terms which need to be defined before proceeding further which are as follows.

- **Feature-Based**: Assembly is created by importing different individual parts, exactly in the same way SOLIDWORKS model is also made up of separate elements which are distinct from others. These elements are called features.

While creating a model using SOLIDWORKS model software, person working should possess sound knowledge, easy to comprehend the nature of the geometric features such as bosses, cuts, holes, ribs, fillets, chamfers, and draft. As features are made, they are implemented directly to the workpiece.

Features can be distinguished into either sketched or applied.

- **Sketched Features**: This type of feature is having a base of 2D sketch. Sketch is transformed into solid by extrusion, rotation, sweeping, or lofting.
- **Applied Features**: This feature is created directly on the solid model. Few examples of applied features are fillets and chamfers.

With the help of the feature manager design tree, SOLIDWORKS software shows the feature-based structure of the model in a special window. Feature manager design tree not only shows the sequence in which the design or model is created but also depicts the inherent associated information. We will be learning about the features in detail throughout the course of SOLIDWORKS and in the upcoming chapters.

- **PARAMETRIC**: The dimensions and relations used to design a feature are captivated and laid in a model. This not only enables you to capture your design intent, it allows you to quickly and easily make changes to the model.
  - **DRIVING DIMENSIONS**: While creating a feature, driving dimensions are used. They include dimensions associated to sketch geometry as well as

feature itself. A simple example of this would be a feature like a cylindrical boss. The diameter of the boss is controlled by the diameter of the sketched circle. The height of the boss is controlled by the depth to which that circle was extruded when the feature was made.

–   **RELATIONS**: These include relations such as parallelism, tangency, and concentricity. Historically, this type of information has been communicated on drawing feature via feature control symbols.

• **SOLID MODELING**: The most necessary part or component used in the CAD systems is known as solid modeling. To determine the edges and faces of the geometric model, it is necessary to have complete wire frames and surface geometry which is included in the solid modeling. In addition to the surface geometry, it also comprises of information known as topology which helps in relating the geometry together. An instance for topology would be surfaces meeting an edge or curve. This intelligence makes operation such as filleting as easy as selecting an edge and specifying its radius.

• **FULLY ASSOCIATIVE**: A model designed in SOLIDWORKS is fully associated with the drawings and assemblies that address it. Changes to the model made are reflected in the drawings and assemblies associated with that particular model. Similarly, changes in the context of the drawing and assembly are reflected back in the model.

• **CONSTRAINS**: Geometric relations such as parallel, perpendicular, horizontal, vertical, concentric, and coincident are some of the constraints used while modeling in SOLIDWORKS software. In addition, equations can be used to establish mathematical relationship among parameters. By using constraints and equations, one can guarantee that design concepts such as through holes or equal radii are captured and maintained.

• **DESIGN INTENT**: Design intent is the plan of a designer as how the model has to behave when it is changed. For example, if a rectangle is modeled with blond hole in it, it should be checked that if the rectangle is moved from its place, the hole also changes its position.

1.21   The SOLIDWORKS User Interface

The SOLIDWORKS user interface is a soul of windows interface and behaves in the same way as other applications of windows. Some of the important aspects of the interface are identified below such as menus, keyboard shortcuts, toolbars, graphics area, feature design tree, quick tips, property manager menus, Task pane, and options.

MENUS: Menu allows way to enter or leave all the commands that are offered by the SOLIDWORKS software. Menus can be seen by clicking on **view**.

As it can be seen that arrow-type symbol appears after display, modify, lights and cameras, etc., which means more options are available in that type.

After clicking on the type, one can see the various sub-types with additional information.

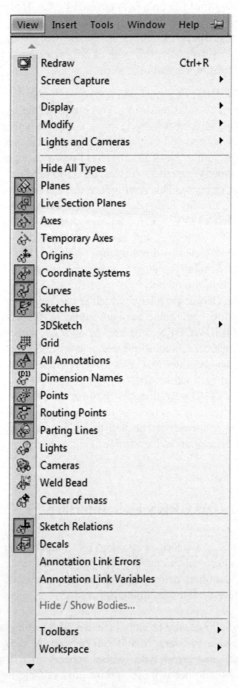

Fig. 1.1 (with permission from Dassault Systems)

When the customize menu is selected as shown in Fig. 1.2, each item appears with the checkbox (Fig. 1.3). Clearing the checkbox removes the associated items from the menu. .

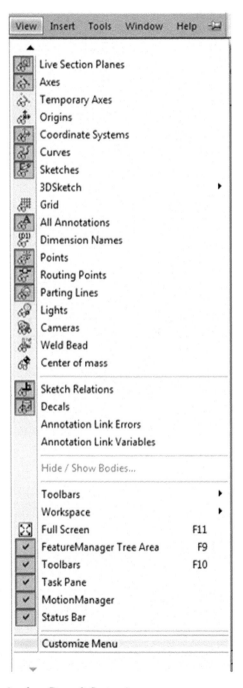

Fig. 1.2 (with permission from Dassault Systems)

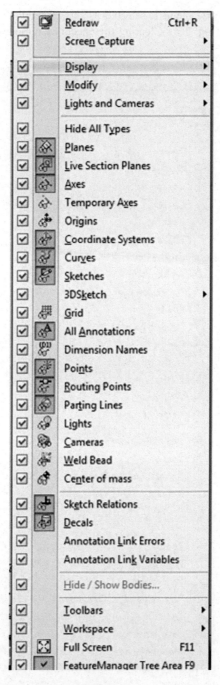

Fig. 1.3 (with permission from Dassault Systems)

KEYBOARD SHORTCUTS: Keyboard shortcuts help the designer in creating his model faster by just knowing the tailor-made shortcuts created by them or by the software itself. In the menu, one can see that redraw can be done just by pressing CTRL+R without clicking on VIEW then on redraw command. SOLIDWORKS adapts to standard windows conventions for such shortcuts as CTRL+S for saving the file, CTRL+O for opening of a new document. CTRL+Z for editing the file, undo, and so on.

TOOLBARS: To quickly approach the most frequently used command, toolbar menus provide shortcut enabling. The toolbars available are organized according to function, and the user can make according to the specifications, remove, or arrange in an order according to their preferences.

Toolbar is shown in the below figure which contains frequently used functions as opening of a new document, opening an existing document, saving the file, printing, copying, editing, undo, and so on.

Fig. 1.4 (with permission from Dassault Systems)

The user/designer can make the toolbars visible by one or two processes:

* Making toolbars visible: click on the **Tools > Customize**.

Fig. 1.5 (with permission from Dassault Systems)

On the toolbars page, click on the checkboxes to select each toolbar you want to display on the graphics window screen. To hide the toolbars, clear the checkboxes.

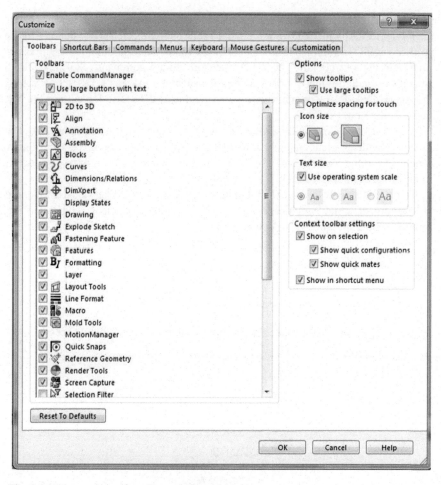

Fig. 1.6 (with permission from Dassault Systems)

In order to access the **Tools, Customize** a document must be opened. Also the **commands** tab can be used to remove and add the icons from toolbars.

Right-clicking on the **options** toolbar opens all the commands and then click on customize present at the bottom. From there also, the toolbars can be made visible and hidden.

QUICK TIPS: Quick tips present in the SOLIDWORKS can be seen at the bottom of the graphics window screen with the symbol of [?]. It is a part of online help system and asks the user whether what you want to do? Based on the work carried out, it gives answers too. Clicking an answer highlights the icons and the toolbars required to perform the task.

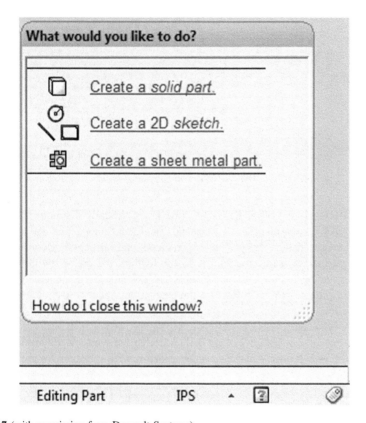

Fig. 1.7 (with permission from Dassault Systems)

When you click on ***create a solid part***, it will start showing the icons which will be used to create a part in SOLIDWORKS.

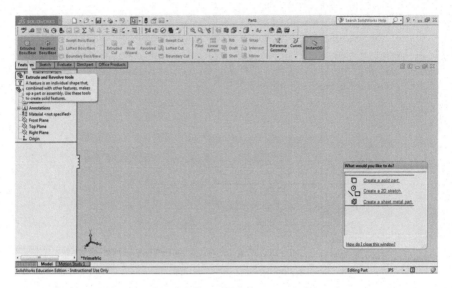

Fig. 1.8 (with permission from Dassault Systems)

FEATURE MANAGER DESIGN TREE: A radically distinctive part of the SOLIDWORKS software that displays all the features in a part or assembly is feature manager design tree. It is represented and arranged according to the sequence maintained while modeling. As the features are created, it gets added to the feature manager, and if some features are deleted, feature manger hides that feature and does not show it. Feature manager allows access to the editing of the feature it contains.

PROPERTY MANAGER MENU: A property manager menu is present next to feature manager design tree in SOLIDWORKS software. Many commands are executed through property manager and replace feature manager when they are in use.

Fig. 1.9 (with permission from Dassault Systems)

TASK PANE: The task pane window is used to put up the SOLIDWORKS **Resources** 🏠 , **Design Library** 📚 , and **File Explorer** 📁 options. This Window appears on the right side of the window screen; it can be resized and moved on the screen.

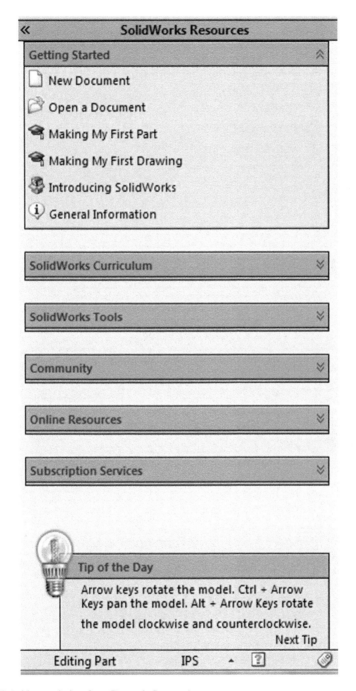

Fig. 1.10 (with permission from Dassault Systems)

Fig. 1.11 (with permission from Dassault Systems)

Fig. 1.12 (with permission from Dassault Systems)

MOUSE BUTTONS: The mouse buttons (left, right, and middle) present in SOLIDWORKS have distinguishable meanings.

- **Left Button**: Select objects such as geometry, menus button, and objects in the feature manager design tree.
- **Right Button**: Make active a context-sensitive shortcut menu. The contents of the menu differ depending on what object the cursor is over. The menus also represent shortcuts to frequently used commands.
- **Middle Button**: In a dynamic manner, it rotates or zooms a part or assembly.

OPTIONS: On the tool menu, one can see the **OPTIONS**. Options customize the SOLIDWORKS software and reflect preferences of the designer and work environment.

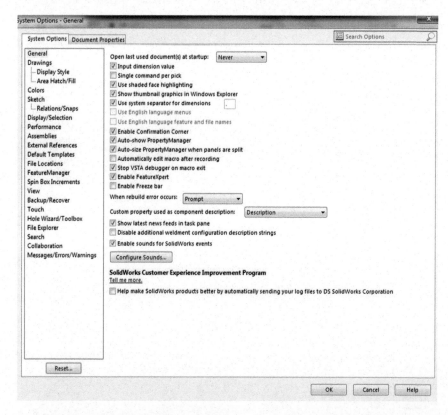

Fig. 1.13 (with permission from Dassault Systems)

CUSTOMIZATION: There are four levels of customization which consists of system options, document properties, document templates, and object.

- **System Options**: The options grouped under the heading system options are kept on the user system and have an effect upon every document user open in his

SOLIDWORKS session. System settings permit the user to assure and customize user's work environment.

- **Document Properties**: Certain settings are applied to the individual document. For instance, units, drafting standards, and material properties (density) are all document setting. They are saved with the document and do not change, regardless of whose system the document is opened on.

- **Document Template**: These are pre-defined documents that are set up with definite specific settings. They can be coordinated into different folders for easy admittance when opening new documents or a file. User can create document templates for parts, assemblies, and drawings.

- **Object**: Many times the properties of an individual object can be changed or edited. For example, you can change the default display of a dimension to suppress one or both extension lines, and you can change the color of a feature.

Chapter 2
Introduction to Sketching

After successful completion of the chapter, the user will be able to design a new part using sketching tools like line, rectangle, circle, parallelogram, arc, and spline. In this chapter, we will be dealing with various types of sketching tools with the help of step-by-step tutorial using figures of the software.

2.1 Line

This command helps in creating a line. To proceed with the line commands, follow the steps as shown below.

1. Click on **New document**.

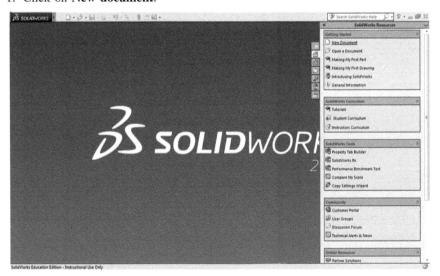

Fig. 2.1 (with permission from Dassault Systems)

© Springer Nature Switzerland AG 2020
K. Kumar et al., *Mastering SolidWorks*, Management and Industrial Engineering,
https://doi.org/10.1007/978-3-030-38901-7_2

2. A tab opens showing **New SOLIDWORKS document**.

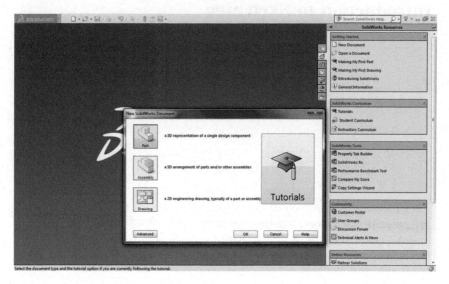

Fig. 2.2 (with permission from Dassault Systems)

3. Select the document type as **Part** > Click **OK**.

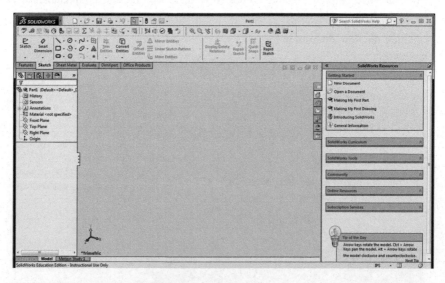

Fig. 2.3 (with permission from Dassault Systems)

4. Go to **Sketch** section and click on **Line** command.

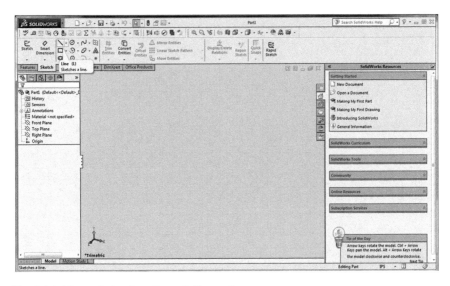

Fig. 2.4 (with permission from Dassault Systems)

5. Select a **PLANE > FRONT PLANE**.

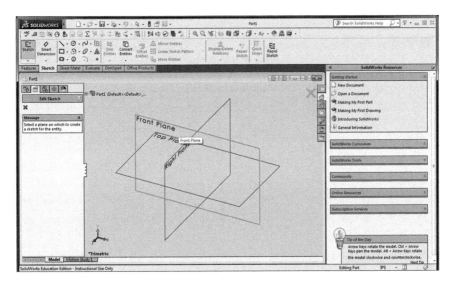

Fig. 2.5 (with permission from Dassault Systems)

6. In the Feature Manager Design Tree, **FRONT PLANE** can be seen. Right-click on **FRONT PLANE** and select **NORMAL TO**.

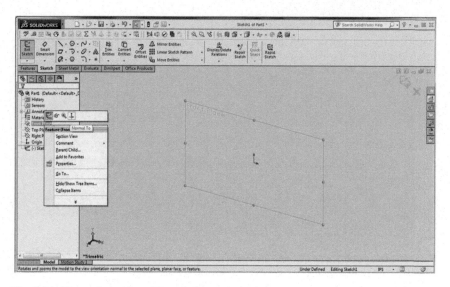

Fig. 2.6 (with permission from Dassault Systems)

7. Activate Line command and create a line by sketching anywhere on the screen. Line can be made horizontal, vertical, as sketched or at any angle.

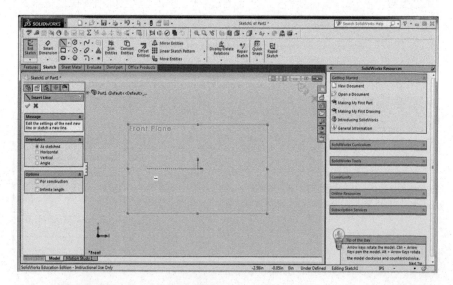

Fig. 2.7 (with permission from Dassault Systems)

8. Create as many lines as specified by clicking on the screen by ending the previous line and start a new one.

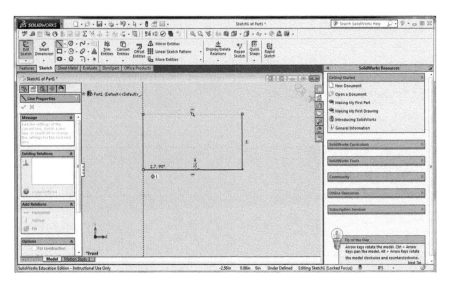

Fig. 2.8 (with permission from Dassault Systems)

9. Double-click to end the command or click **OK**.

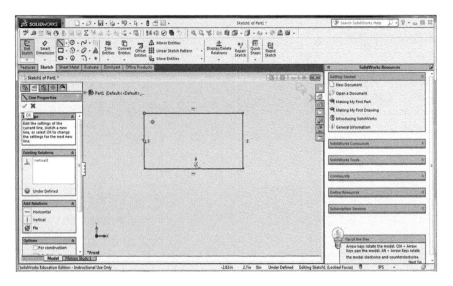

Fig. 2.9 (with permission from Dassault Systems)

2.1.1   *Centerline*

This command is used to create a centerline. Centerline can be used to create symmetrical sketch elements, revolved features or as construction of geometry. It can be created as shown below.

1. Click on the CENTERLINE.

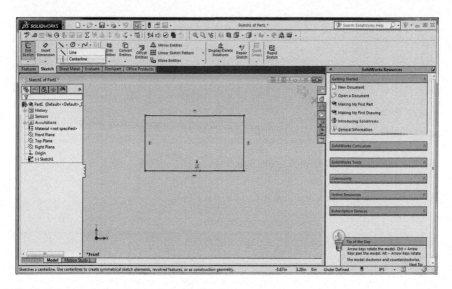

Fig. 2.10 (with permission from Dassault Systems)

2. Create a centerline from one end to another and double-click on the screen to end the command.

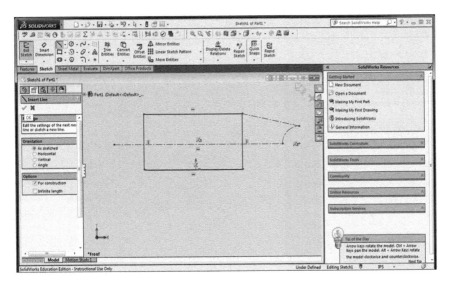

Fig. 2.11 (with permission from Dassault Systems)

2.2   Rectangle

There are five types of rectangle discussed in SOLIDWORKS. Here, we will be discussing each one by one.

2.2.1   Corner Rectangle

To start with corner rectangle

1. Go to **Sketch**.
2. Select **Corner Rectangle**.
3. Click on **Plane** to select **Front plane**.

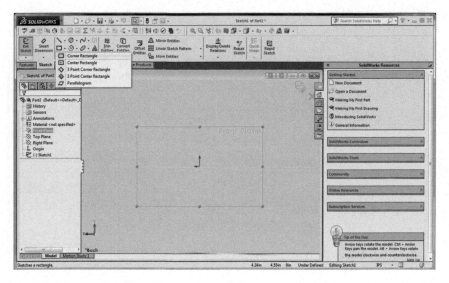

Fig. 2.12 (with permission from Dassault Systems)

4. Click on the screen and drag to create a **corner rectangle**.
5. Click **OK** to complete the **Corner Rectangle**.

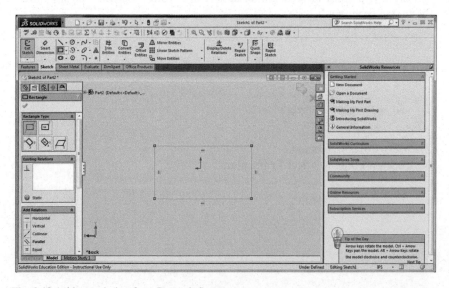

Fig. 2.13 (with permission from Dassault Systems)

2.2.2 Center Rectangle

This type of rectangle sketches a rectangle from the center. To proceed with this type of rectangle, follow the steps as given below.

1. Select the plane.

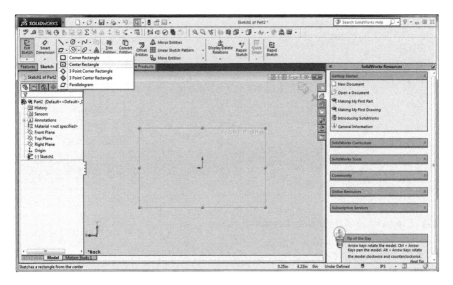

Fig. 2.14 (with permission from Dassault Systems)

2. Select the rectangle > Center Rectangle.

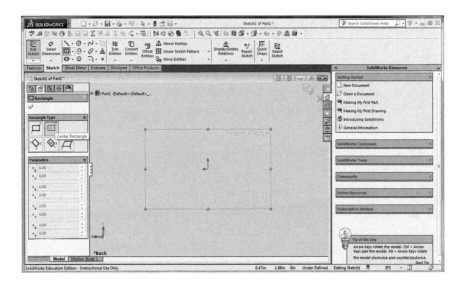

Fig. 2.15 (with permission from Dassault Systems)

3. Click on the center and with the help of mouse drag the rectangle.
4. Click **OK** to complete the rectangle.

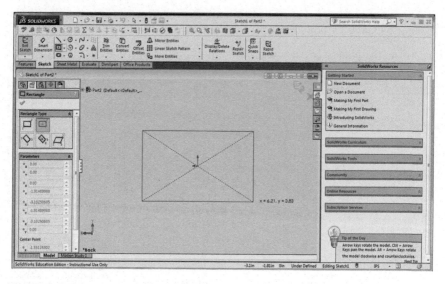

Fig. 2.16 (with permission from Dassault Systems)

2.2.3   3 Point Corner Rectangle

This type of rectangle sketches a rectangle with an angle. To start with 3 Point Corner Rectangle

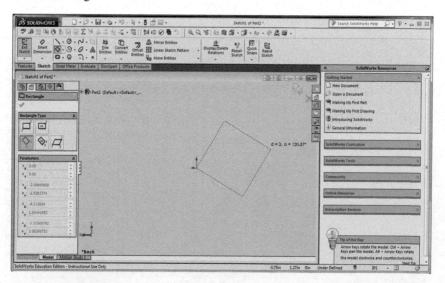

Fig. 2.17 (with permission from Dassault Systems)

1. Click on the command from the Rectangle section of the sketch.
2. Select the plane.
3. Click anywhere on the screen and start sketching rectangle with an angle as shown in the figure.

2.2.4 3 Point Center Rectangle

This type of rectangle sketches an angle rectangle from the center. To sketch, follow the given steps.

1. Select a plane on which to create a sketch.
2. Click center, width, and the length.
3. Click **OK** to complete the command.

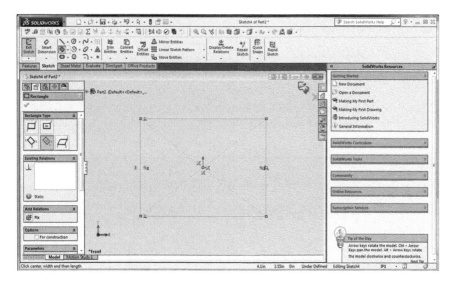

Fig. 2.18 (with permission from Dassault Systems)

2.2.5 Parallelogram

Sketches a parallelogram

1. Select the plane on which sketch is to be created.
2. Click first, second, and then third corner to complete.
3. Click **OK** to end the command.

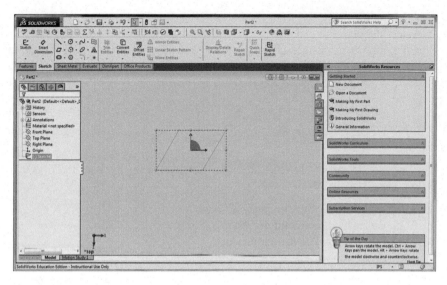

Fig. 2.19 (with permission from Dassault Systems)

2.3   Slot

Slot can be of four types such as Straight Slot, Centerpoint Straight Slot, 3 Point Arc Slot and Centerpoint Arc Slot.

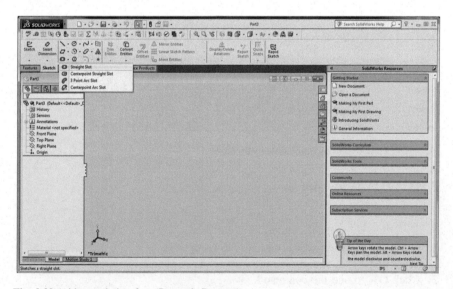

Fig. 2.20 (with permission from Dassault Systems)

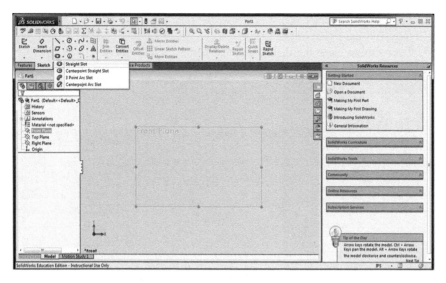

Fig. 2.21 (with permission from Dassault Systems)

2.3.1   Straight Slot

This type of slot sketches a straight slot. This can be created with the help of given below steps.

1. Select the plane on which to create a sketch for the entity.
2. Click on Straight Slot.
3. Click on the graphics window screen and draw a line as shown in the figure.

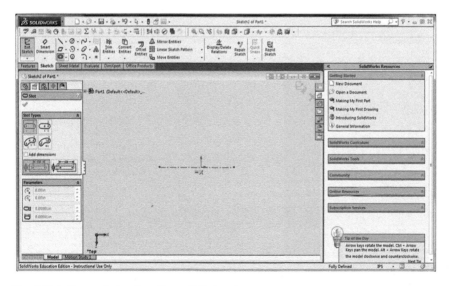

Fig. 2.22 (with permission from Dassault Systems)

4. Then drag the cursor above or below the line; a straight slot will be created as shown in the figure below.

5. Click **OK** to close the dialog box and complete the slot.

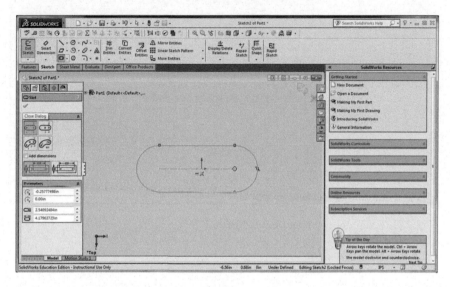

Fig. 2.23 (with permission from Dassault Systems)

2.3.2 Centerpoint Straight Slot

Sketches a Centerpoint Straight Slot. To create centerpoint straight slot, select the plane and click on the center anywhere on the graphics window screen. Drag the cursor, a centerpoint straight slot will be created.

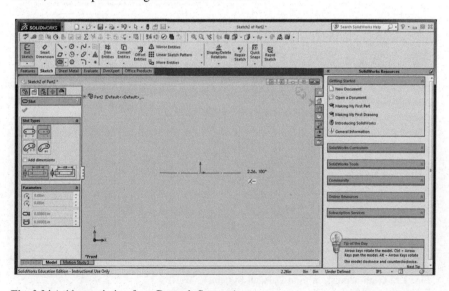

Fig. 2.24 (with permission from Dassault Systems)

Click **OK** to complete the feature and end the command.

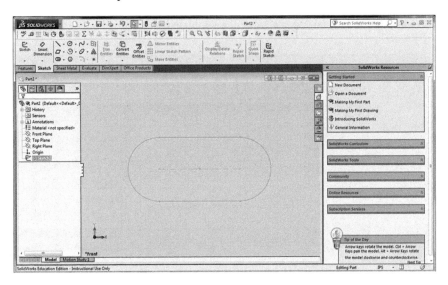

Fig. 2.25 (with permission from Dassault Systems)

2.3.3   3 Point Arc Slot

It creates a 3 Point Arc Slot. To start with this type of slot, follow the steps one by one as given below.

1. Select a plane.
2. Click on 3 Point Arc Slot.
3. Click on the first, second and third point to create an Arc.

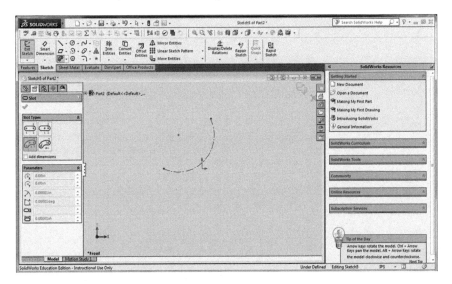

Fig. 2.26 (with permission from Dassault Systems)

4. Drag the mouse cursor from the arc to different position to create 3 Point Arc Slot.
5. Close the dialog box to complete the command.

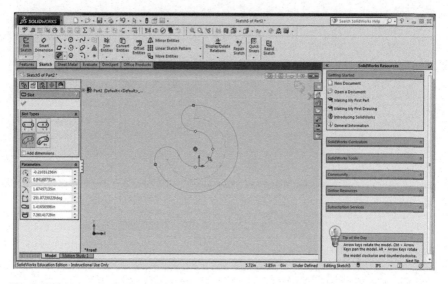

Fig. 2.27 (with permission from Dassault Systems)

2.3.4 Centerpoint Arc Slot

It sketches a Centerpoint Arc Slot.

1. Click on the command.

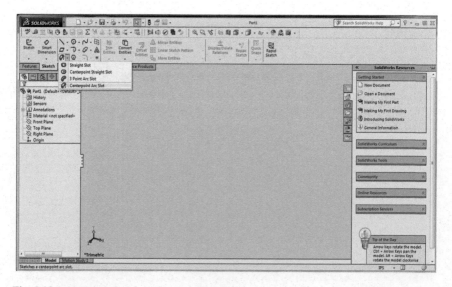

Fig. 2.28 (with permission from Dassault Systems)

3. Select the plane.
4. Click on the center and drag the cursor to create a 3 Point Arc Slot.

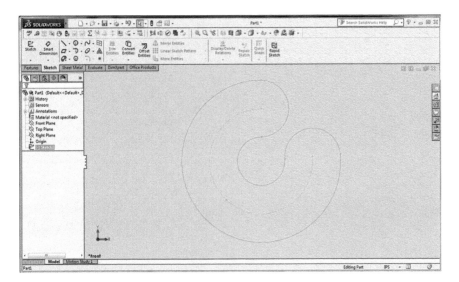

Fig. 2.29 (with permission from Dassault Systems)

2.4 Circle

Circle command is used to create a circle. In SOLIDWORKS, circle and perimeter circle are present which will be discussed with the help of the steps followed and the figures.

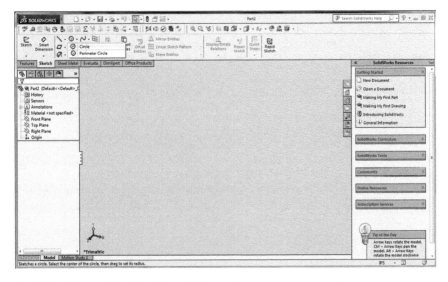

Fig. 2.30 (with permission from Dassault Systems)

To create a circle, select the center of the circle and then drag to set its radius. A circle will be created as shown below in the figure.

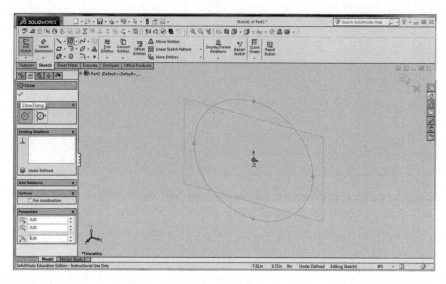

Fig. 2.31 (with permission from Dassault Systems)

Perimeter Circle: Sketches a circle by its perimeter. Select a point on the perimeter, then a second and a third (optionally). A perimeter circle will be created.

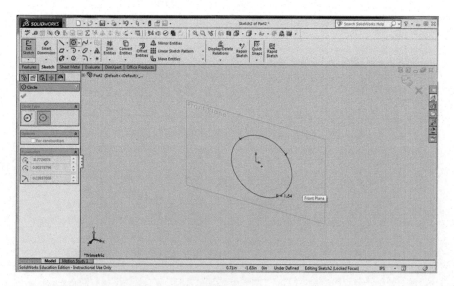

Fig. 2.32 (with permission from Dassault Systems)

In the feature design tree, right-click on the sketch and then select Normal to. The part will be displayed on the screen as shown in the figure.

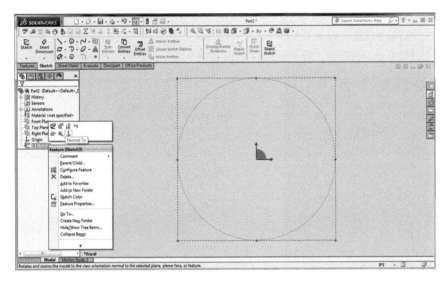

Fig. 2.33 (with permission from Dassault Systems)

2.5 ARC

Three different types of arc can be created in SOLIDWORKS software such as Centerpoint Arc, Tangent Arc and 3 Point Arc.

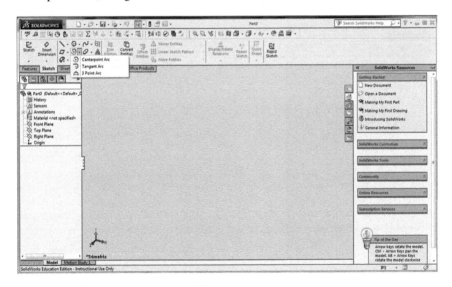

Fig. 2.34 (with permission from Dassault Systems)

2.5.1 Centerpoint Arc

Sketches a centerpoint Arc.

2. Set the centerpoint.
3. Drag to place the Arc starting point, then to set its length and direction.

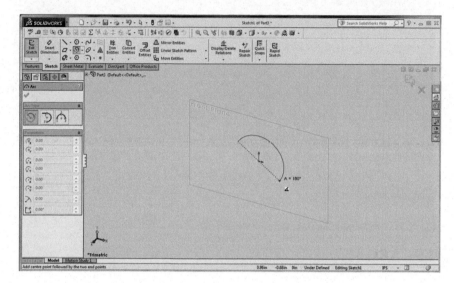

Fig. 2.35 (with permission from Dassault Systems)

4. Double-click to end the Arc.

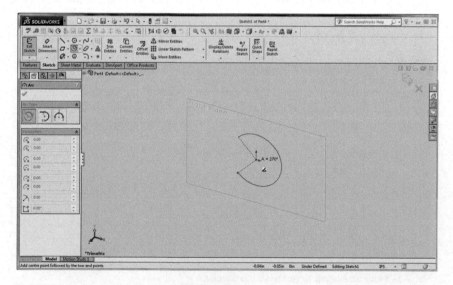

Fig. 2.36 (with permission from Dassault Systems)

5. Click **OK** to close the dialog box and end the feature.

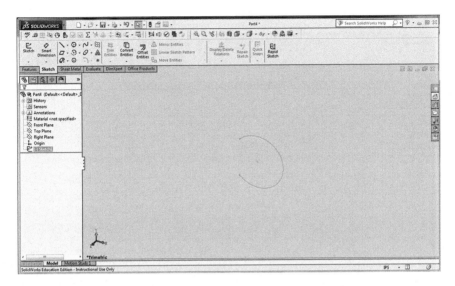

Fig. 2.37 (with permission from Dassault Systems)

2.5.2   *Tangent Arc*

Sketches an Arc tangent to sketch an entity.

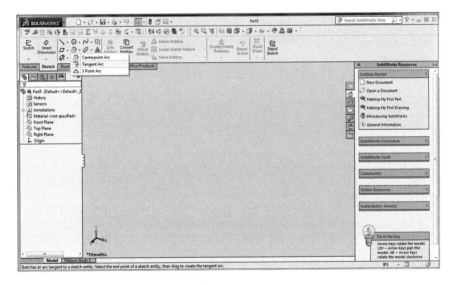

Fig. 2.38 (with permission from Dassault Systems)

1. Select a plane to create a sketch.
2. Click on Line command and create two lines.

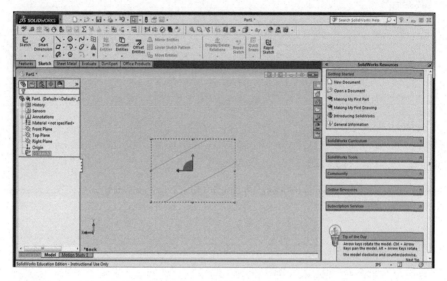

Fig. 2.39

3. Select the endpoint of a sketch entity.

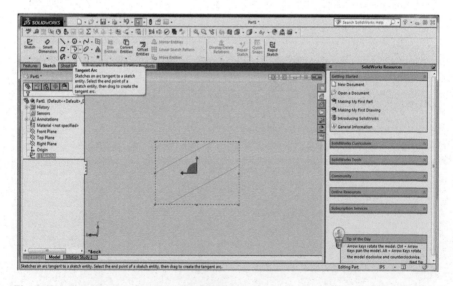

Fig. 2.40 (with permission from Dassault Systems)

4. Join the endpoints as shown and then drag to create a tangent arc.

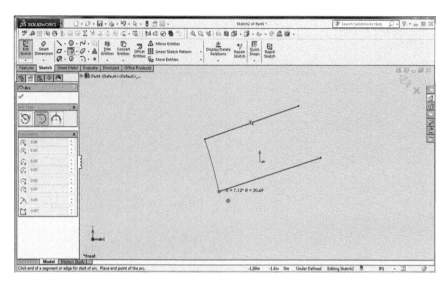

Fig. 2.41 (with permission from Dassault Systems)

2.5.3 3 Point Arc

Sketches a 3 Point arc. Select the endpoint and the starting points and then drag the arc to set the radius or to reverse the arc.

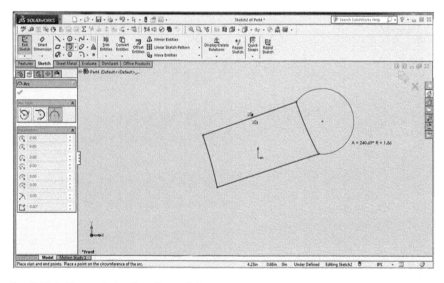

Fig. 2.42 (with permission from Dassault Systems)

2.6 Polygon

It sketches a polygon. The user can change the number of sides after sketching the polygon.

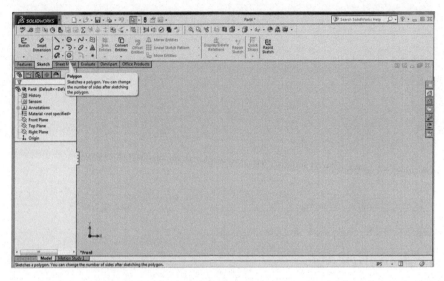

Fig. 2.43 (with permission from Dassault Systems)

Set numbers of sides then click and drag to create a polygon.

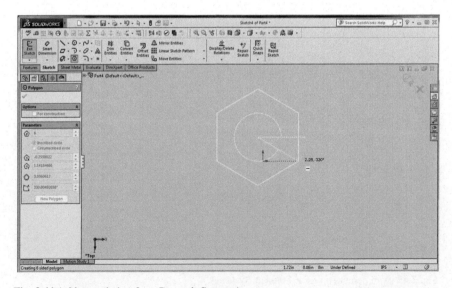

Fig. 2.44 (with permission from Dassault Systems)

Close the dialog box to complete the polygon.

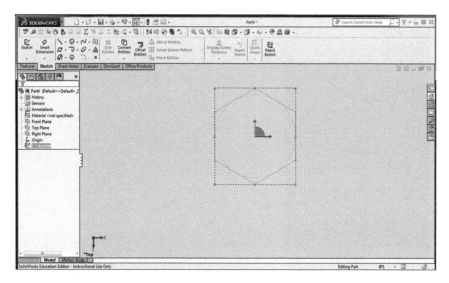

Fig. 2.45 (with permission from Dassault Systems)

2.7 Spline

Spline command helps in sketching a spline. It can be created in four different types such as spline, style spline, spline on surface and equation driven curve.

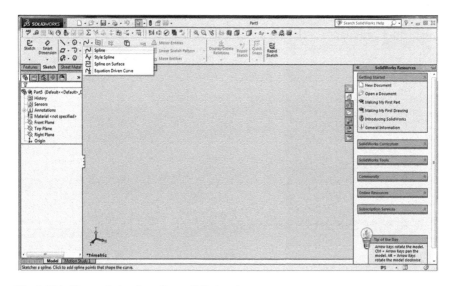

Fig. 2.46 (with permission from Dassault Systems)

Select the plane and click to add spline points that shape the curve.

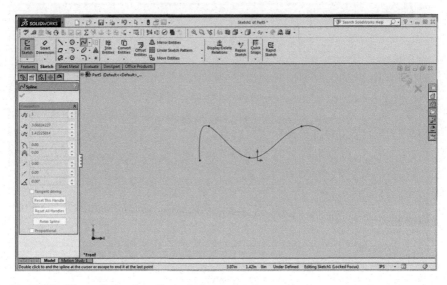

Fig. 2.47 (with permission from Dassault Systems)

Click **OK** to complete the Spline feature.

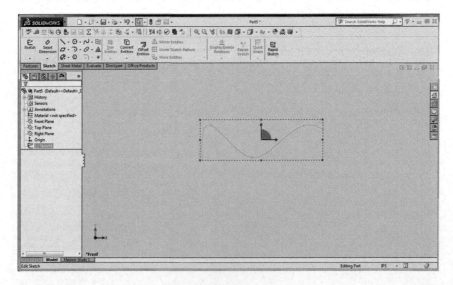

Fig. 2.48 (with permission from Dassault Systems)

Double-click to end the spline at the cursor or escape to end the last point.

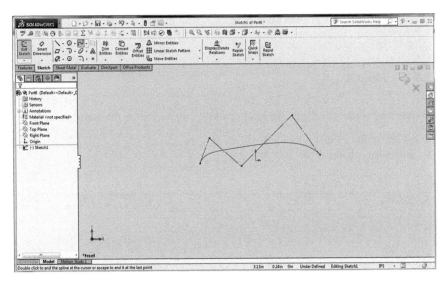

Fig. 2.49 (with permission from Dassault Systems)

Click on the **Style Spline** and adjust the scale of selected sketch entity.

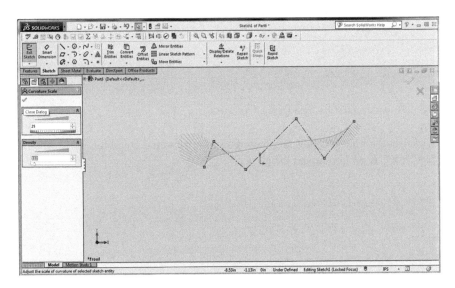

Fig. 2.50 (with permission from Dassault Systems)

2.8   Ellipse

Ellipse is distributed in four parts ellipse, partial ellipse, parabola and conic.

2.8.1   Ellipse

Sketches a complete ellipse. To start creating ellipse follow the steps stated below.

1. Click on the ellipse
2. Select the plane.

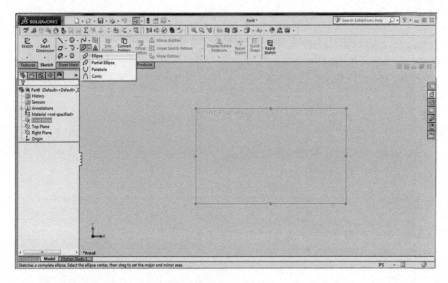

Fig. 2.51 (with permission from Dassault Systems)

3. Select the ellipse center and then drag to set the major and the minor axes.
5. Click **OK** and complete the ellipse.

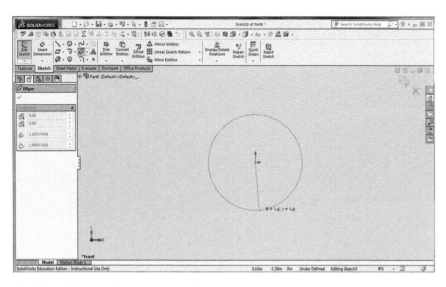

Fig. 2.52 (with permission from Dassault Systems)

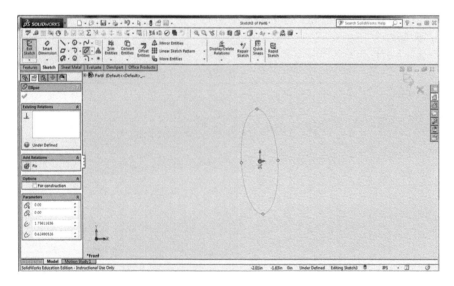

Fig. 2.53 (with permission from Dassault Systems)

2.8.2   Partial Ellipse

Click on the Partial Ellipse command to sketch a partial ellipse.

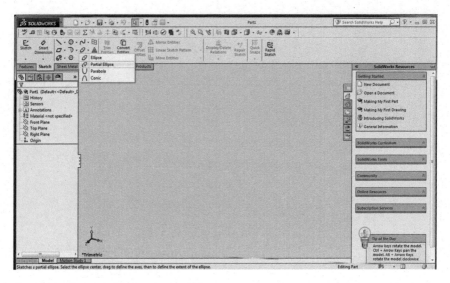

Fig. 2.54 (with permission from Dassault Systems)

6. Select the ellipse center, drag to define the axes, then to define the extent of the ellipse.
7. Click **OK** to complete the Partial Ellipse.

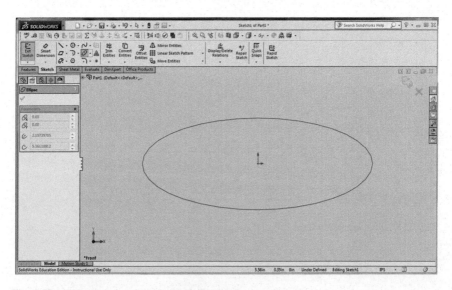

Fig. 2.55 (with permission from Dassault Systems)

8. Drag the cursor on the ellipse drawn and then edit it using different parameters.

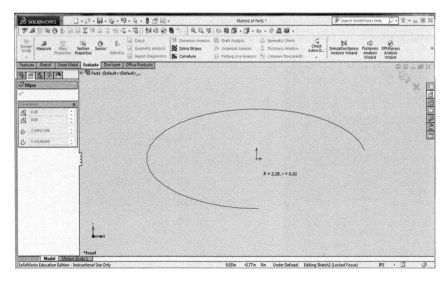

Fig. 2.56 (with permission from Dassault Systems)

2.8.3   Parabola

Sketches a parabola. Click on the command.

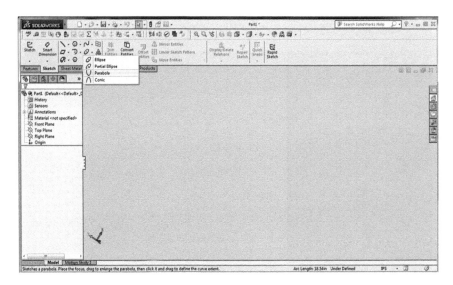

Fig. 2.57 (with permission from Dassault Systems)

Select the plane. Place the focus, drag to enlarge the parabola, then click it and drag to define the curve extent. Click **OK** to complete parabola.

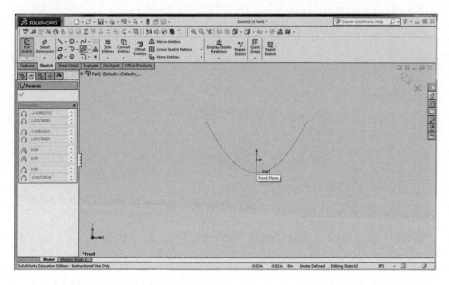

Fig. 2.58 (with permission from Dassault Systems)

2.8.4 Conic

This command sketches a conic. Select the conic command from the sketch section.

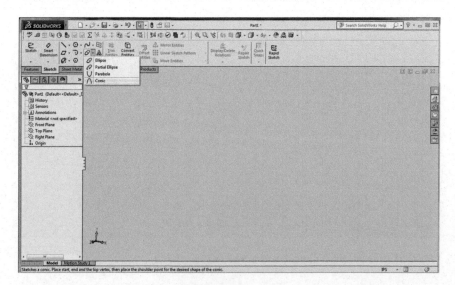

Fig. 2.59 (with permission from Dassault Systems)

Place start, end and the top vertex.

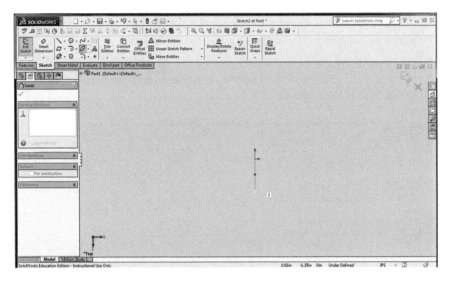

Fig. 2.60 (with permission from Dassault Systems)

Then place the shoulder point for the desired shape of the conic. Click **OK** to complete the command.

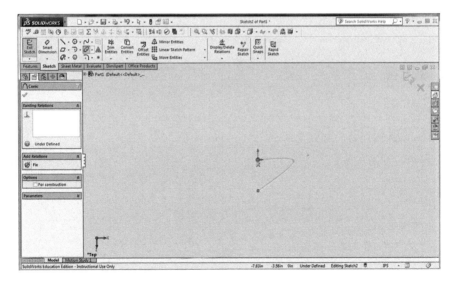

Fig. 2.61 (with permission from Dassault Systems)

2.9   Sketch Fillet

Rounds the corner at the intersection of two sketched entities, creating a tangent arc. To start with fillet, the user needs to follow the given steps.

1. Select the Plane
2. Click on Line command and create a rectangle with the help of it.
3. Select Sketch Fillet option.

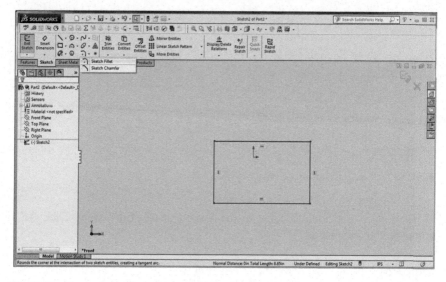

Fig. 2.62 (with permission from Dassault Systems)

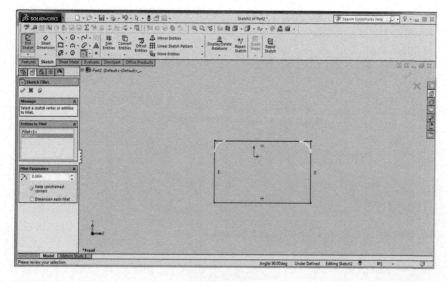

Fig. 2.63 (with permission from Dassault Systems)

4. Select the first entity to fillet and then press CTRL to select second entity to fillet as shown in the figure.

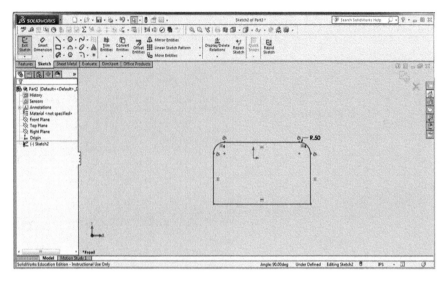

Fig. 2.64 (with permission from Dassault Systems)

5. Review the selection.
6. Click **OK** to complete feature.

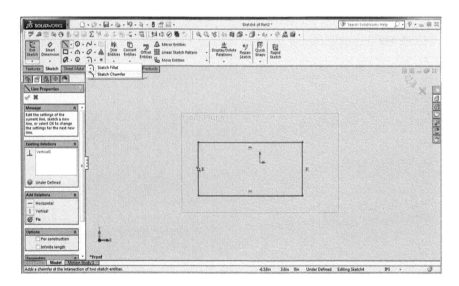

Fig. 2.65 (with permission from Dassault Systems)

2.10   Sketch Chamfer

Adds a chamfer at the intersection of two sketched entities. To know about chamfer and how it is created draw rectangle with the help of lines as shown in the figure and select Sketch Chamfer.

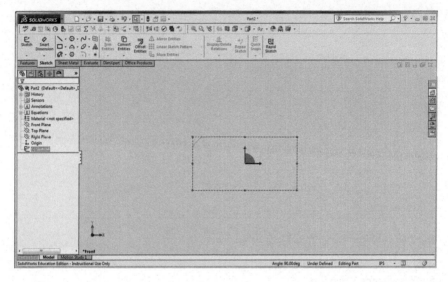

Fig. 2.66 (with permission from Dassault Systems)

Select a corner or first entity to chamfer and then select the second one. Review the selection.

Click **OK** to complete the feature.

Chapter 3
Basic Sketch Relations and Dimensioning

This chapter will cover the basic automatic sketch relations, sketch feedback, turning off tools which will also include the status of the sketch while designing a part or a model. It will also include examples of sketch relations and rules that govern the sketch designs.

We will be dealing with each of them one by one with the help of examples and figures with step-by-step approach to make it clear to the user effectively.

3.1 Inference Lines (Automatic Relations)

In addition to the horizontal and vertical lines symbols, dashed inference lines will also appear to help the designer 'line up' with the existing geometry. These lines include line vectors, normal, horizontals, verticals, tangents, and centers. It should be noted that some lines capture actual geometric relations while other simply acts as a guide or reference when sketching. A difference in the color of the inference lines will distinguish them.

The display of the inference lines (Sketch Relations) that appear automatically can be toggled on and off using **view, Sketch Relations**. They will remain on during the initial phase of sketching.

© Springer Nature Switzerland AG 2020
K. Kumar et al., *Mastering SolidWorks*, Management and Industrial Engineering,
https://doi.org/10.1007/978-3-030-38901-7_3

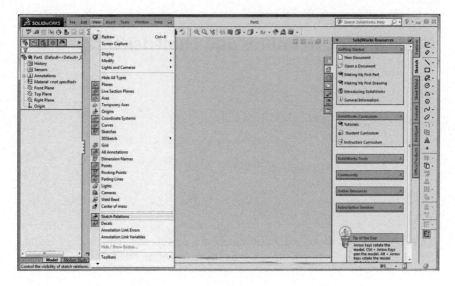

Fig. 3.1 (with permission from Dassault Systems)

Moving in a direction perpendicular to the previous line causes inference lines to be displayed. A **perpendicular** relation is created using this line and the last one.

The cursor symbol indicates that you are capturing a perpendicular relation. Note that the line cursor is not shown for clarity.

3.1.1   Perpendicular

Another perpendicular line is created from the last endpoint. Again, a perpendicular relation is automatically captured.

3.1.2   Reference

Some inferences are only for references and do not create relations. They are displayed in blue color. This reference is used to align the endpoint vertically with the origin.

3.2   Sketch Feedback

The sketcher has many feedback features. The cursor will change to show what type of entity is being created. It will also indicate what selections on the existing geometry such as end, coincident, (on) or midpoint is available using a red dot when the cursor is on it.

Three of the most common feedback symbols are:

3.2.1   *Endpoint*

Yellow concentric circles appear at the midpoint when the cursor is over it.

3.2.2   *Midpoint*

The midpoint appears as a square. It changes to red when the cursor is over the line.

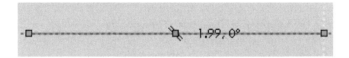

3.2.3   *Coincident*

The quadrants point of the circle appears with the concentric circle over the center point.

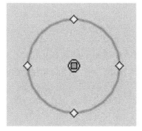

3.3   Close

Close the sketch with a final line connected to the starting point of the first line.

3.4   Turning off Tools

Turn off the active tool can be done using one of the below given techniques.

1. By pressing the **ESC** key on the keyboard.
2. Clicking the tool a second time that is y double-clicking on the tool.
3. By clicking on the **select** (arrow) tool.
4. Right-click in the graphics area, select from the shortcut menu.

3.5   Status of the Sketch

The status of the sketch depends upon the geometric relations between the geometry and the dimensions that define it.

1. **Under Defined**: There is inadequate definition of the sketch, but the sketch can be still be used to create features. This is good because many times in the early process, there is not sufficient information to fully define the sketch created. When more information becomes available, the remaining definition can be added a later time. Under defined sketch geometry is blue (by default).
2. **Fully Defined**: The sketch has complete information. Fully defined geometry is black (by default). As a general rule, when a part is released for manufacturing, the sketches within it should be fully defined.
3. **Over Defined**: The sketch has duplicate dimensions or conflicting relations, and it should not be used until repaired. Extraneous dimensions and definitions must be deleted. Over defined geometry is red in color (by default).

3.6   Design Intent

The design intent governs how the part is built and how will it change. In this given below examples, it is shown that how a sketch created can be changed into another shape.

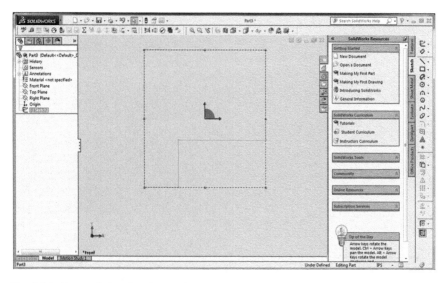

Fig. 3.2 (with permission from Dassault Systems)

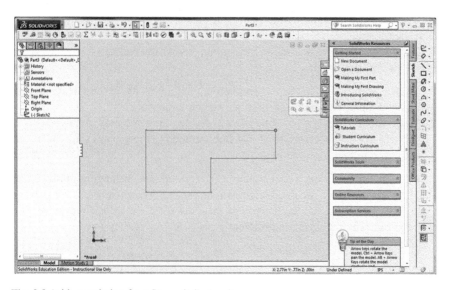

Fig. 3.3 (with permission from Dassault Systems)

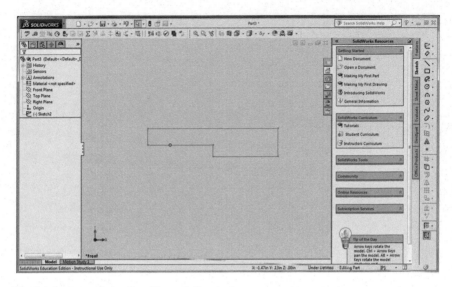

Fig. 3.4 (with permission from Dassault Systems)

By dragging the points of the sketch the shape can be changed which is known as design intent.

Design intent can be controlled by a combination of two things which are **sketch relations** and **dimensions**.

1. **Sketch Relations**: It creates geometric relations such as collinear, parallel, perpendicular, or coincident between sketch elements.
2. **Dimensions**: They are basically used to define the sketch may giving size to the part created and location of the sketch geometry. Linear, angular, radial, and diametric dimensions can be added.

To fully define a sketch and capture the required design intent requires understanding and applying a combination of relations and dimensions.

In order for the sketch to change properly, the correct relations and dimensions are required. The required design intent is listed below one by one

1. Horizontal and vertical lines
2. Angle value
3. Parallel distance value
4. Right angle corners or perpendicular lines
5. Overall length value.

Sketch Relations: are used to force a behavior on a sketch element thereby capturing design intent. Some are automatic; others can be added as needed.

Automatic Sketch Relations: are added as geometry is sketched. We saw this as we sketched the outline in the previous steps. Sketch feedback tells you when automatic relations are being created.

Added Sketch Relations: For those relations that cannot be added automatically, tool exists to create relations based on selected geometry and add dimensions.

Display Relations: It shows and optionally allows you to remove geometric relationships based on selected geometry and add dimensions.

To find the display relations all you have to do is to.

1. Double-click the entity: Symbols appear indicating what relations are associated with entity. In the below example, it is seen that existing relation shown is horizontal.

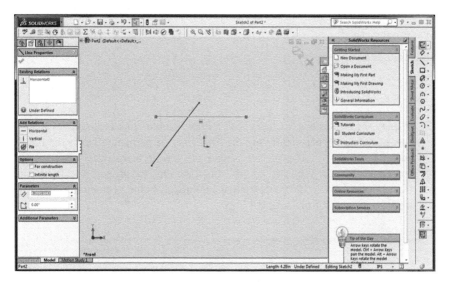

Fig. 3.5 (with permission from Dassault Systems)

2. The property Manager: Select the sketch entity and the property manager shows the relations associated with that entity. Click on the Display/Delete relations.

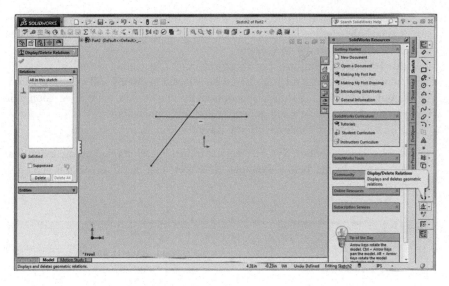

Fig. 3.6 (with permission from Dassault Systems)

3. Display the relations associated with a line: Double-click the uppermost angled line. Symbols appear identifying the lines that are perpendicular to the line you selected.
4. Property Manager: When you double-click the line, the property manager opens. The relations box in the property manager also lists the geometric relations that are associated with the selected line.

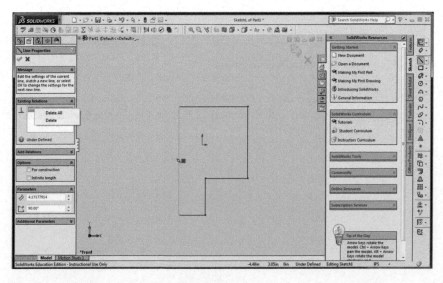

Fig. 3.7 (with permission from Dassault Systems)

5. Remove the relation: Remove the uppermost relation by clicking the relation, either the symbol or in the property manager, and pressing the Delete key. If the symbol is selected, it turns yellow and displays the entities it controls.
6. Drag the endpoint: Because the line is no longer constrained to be perpendicular, the sketch will behave differently when you drag it.

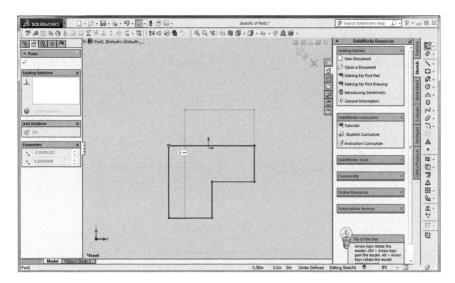

Fig. 3.8 (with permission from Dassault Systems)

Examples of Sketch Relations: There are many types of sketch relations. The validity of the one selected depends upon the combination of the geometry that you select. Selections can be entity itself, endpoints or combination. Depending on the selection, a limited set of options is made available. The following chart shows some examples of sketch relations. This is not a complete list of all geometric relations. Additional examples will be introduced throughout this course.

1. **Coincident**: Coincident between the line and the endpoint. By pressing CTRL select the line and the endpoint of another line. **Add relations** tab will open on the left side of the screen. Select **coincident**.

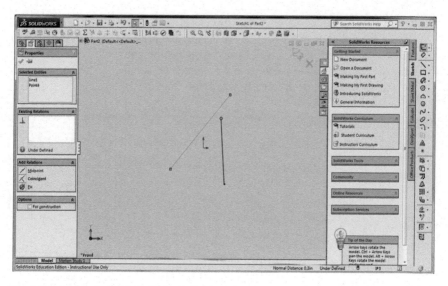

Fig. 3.9 (with permission from Dassault Systems)

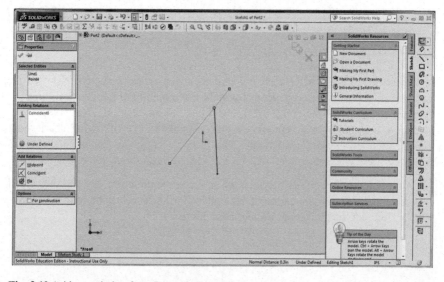

Fig. 3.10 (with permission from Dassault Systems)

2. **Merge**: It is between two endpoints. To merge two lines by pressing **CTRL** select two endpoints of the line. **Add relations** as **Merge** and it is seen that both the lines will merge into one.

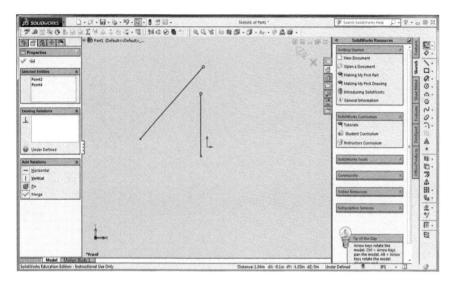

Fig. 3.11 (with permission from Dassault Systems)

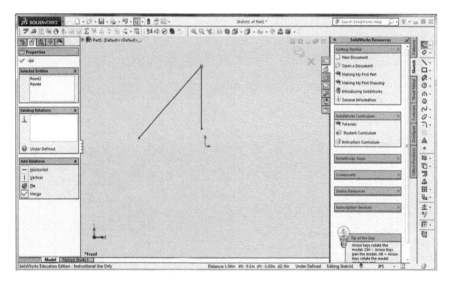

Fig. 3.12 (with permission from Dassault Systems)

3. **Parallel**: This is done between two lines. Draw two lines and by pressing **CTRL** select both the lines one by one. **Add relations** as **Parallel**.

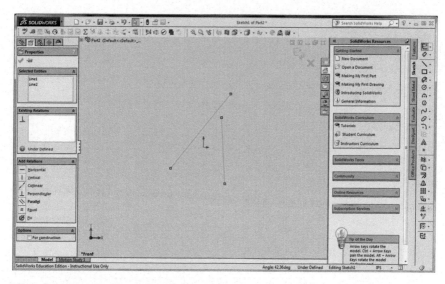

Fig. 3.13 (with permission from Dassault Systems)

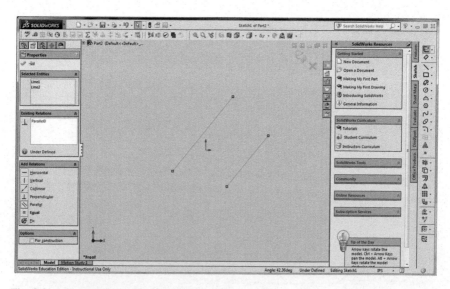

Fig. 3.14 (with permission from Dassault Systems)

4. **Perpendicular**: Between two lines. With the help of the same sketch as shown in the previous above examples **add relations** by selecting both of the lines. Click on **Perpendicular**.

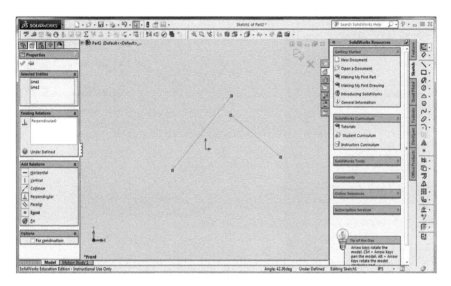

Fig. 3.15 (with permission from Dassault Systems)

5. **Collinear:** Between two lines. Select two lines by sketching it with the help of line command. Select line 1 and line 2. **Add relations** as **Collinear**.

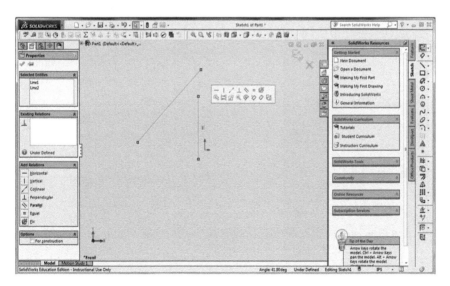

Fig. 3.16 (with permission from Dassault Systems)

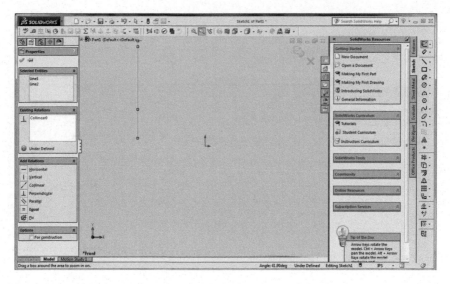

Fig. 3.17 (with permission from Dassault Systems)

6. **Horizontal**: It is applied to one or more lines. With the help of CTRL button select one or more lines and add relations as **Horizontal**.

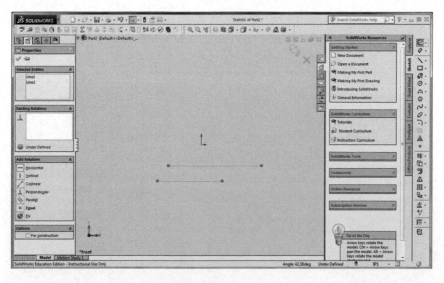

Fig. 3.18 (with permission from Dassault Systems)

7. **Horizontal**: Between two endpoints. Select the endpoints of the lines one by one by pressing CTRL. Before and after changes can be viewed from the figure shown below.

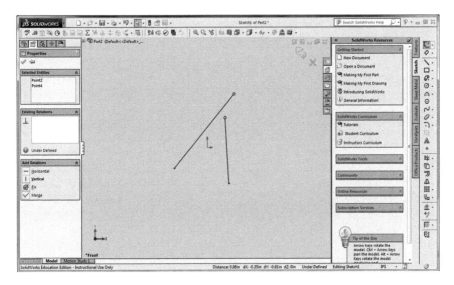

Fig. 3.19 (with permission from Dassault Systems)

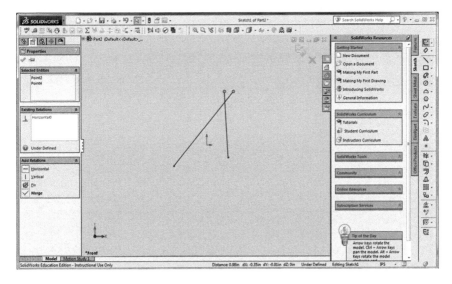

Fig. 3.20 (with permission from Dassault Systems)

8. **Vertical**: Applied to one or more lines. It is done in the same way as done with the horizontal type and the difference between the figure shows the relations importance.

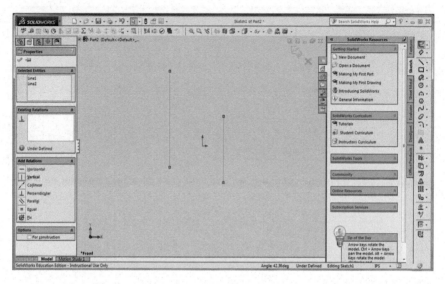

Fig. 3.21 (with permission from Dassault Systems)

9. **Vertical**: Between two endpoints. Selection of point 1 and point 2 by pressing CTRL and then adding relation as **Vertical**.

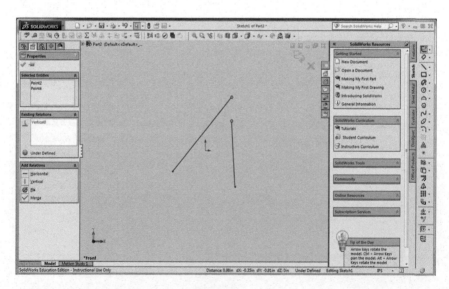

Fig. 3.22 (with permission from Dassault Systems)

10. **Equal**: Between lines and two arcs and circles. Select the line 1 and line 2 with unequal dimension and **add relations** as **Equal**. The equal relation will make both the lines equal in size as per the dimensions as can be seen from the figure.

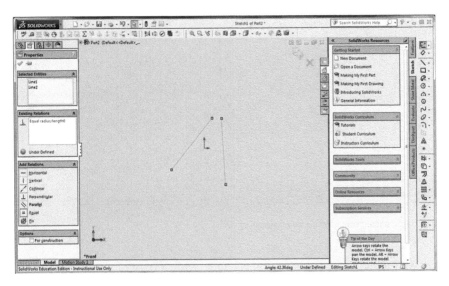

Fig. 3.23 (with permission from Dassault Systems)

Go to sketch and click on circle tool. Draw two circles or arcs of different dimension. Now select both the circle one by one by pressing **CTRL** button. **Add relations** as **Equal**. The equal relations will change make the other circle also of same radius to make it equal.

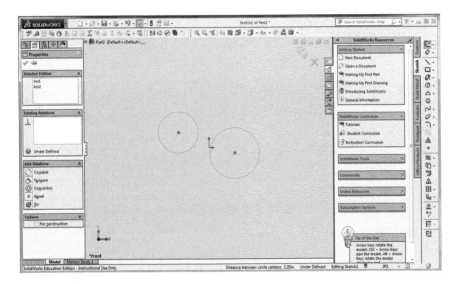

Fig. 3.24 (with permission from Dassault Systems)

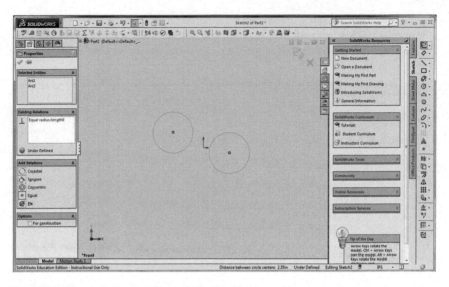

Fig. 3.25 (with permission from Dassault Systems)

11. **Midpoint**: Between line and an endpoint. Draw two lines. Select both the lines one by one by pressing **CTRL**. **Add relations** as **Midpoint**. The other line will be placed at the midpoint of line 1 as can be seen from the figure.

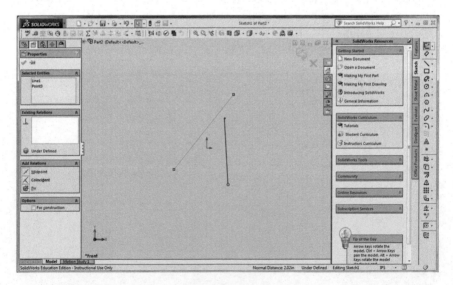

Fig. 3.26 (with permission from Dassault Systems)

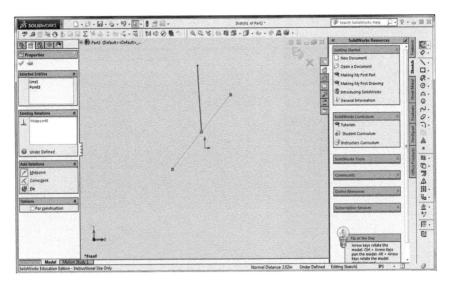

Fig. 3.27 (with permission from Dassault Systems)

3.7 Dimensions

Dimensions are another way to define geometry and capture design intent in the SOLIDWORKS software. The advantage of using dimensions is that it is used to display the current value and change it.

Introduction to Smart Dimensions: The Smart Dimension tool determines the proper type of dimension based on the geometry chosen, previewing the dimension before creating it. For example, if you pick an arc, the system will create a radial dimension. If you pick a circle, you will get a diameter dimension, while selecting two parallel lines will create a linear dimension between them. In cases where the Smart Dimension tool is not quite smart enough, you have the option of selecting endpoints and moving the dimension to different measurement positions.

1. It can be selected by clicking on the **Tools menu**, Select **Dimensions**, **Smart**.

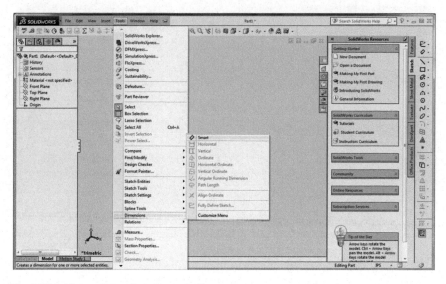

Fig. 3.28 (with permission from Dassault Systems)

2. Right-click and select **smart dimensions** from the shortcut menu.

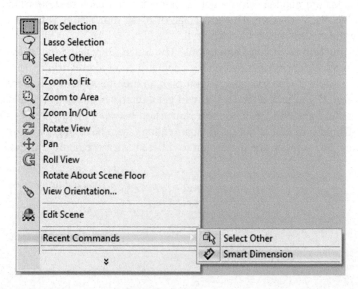

Fig. 3.29 (with permission from Dassault Systems)

3. On the **Dimensions/Relations toolbar**, click on the **Smart Dimension** tool .

Dimensioning Selection and Preview: As you select sketch geometry with the dimension tool, the system creates a preview of the dimension. The preview allows you to see all the possible options by simply moving the mouse after making the selections. Clicking the left mouse button places the dimension in its current position and orientation. Clicking the right mouse button locks only the orientation, allowing you to move the text before final placement by clicking the left mouse button.

With the dimension tool and two endpoints selected, below are three possible orientations for a linear dimension. The value is derived from the initial point to point distance and may change based on the orientation selected.

Create a sketch to add dimensions. Click on the endpoints of the line and move the cursor upwards. Length of 3.82 can be seen and modify dialog box appears.

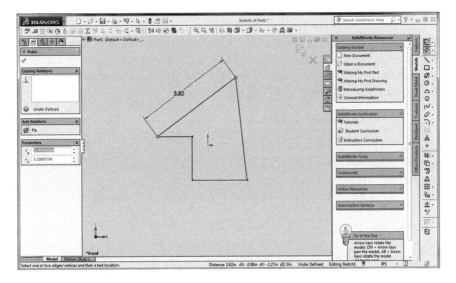

Fig. 3.30 (with permission from Dassault Systems)

Edit the dimension from 3.82 to 3 inch. Units can also be changed as per the requirement of the designer.

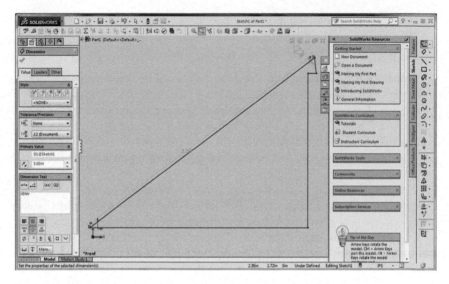

Fig. 3.31 (with permission from Dassault Systems)

Adding a Linear Dimension: Select the dimension tool from any source and click on the line shown in the figure. Click a second time to place the text of the dimension above and to the right of the line. The dimension appears with a Modify tool displaying the current length of the line.

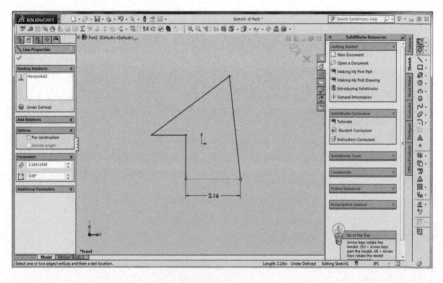

Fig. 3.32 (with permission from Dassault Systems)

Edit the dimension by double-clicking on it and enter the dimension in the dialog box appearing (Modify tool).

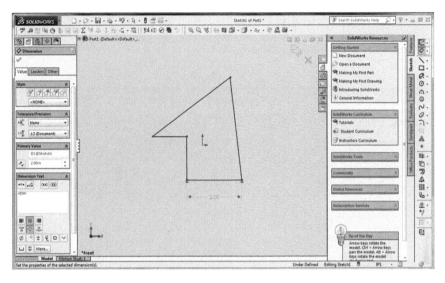

Fig. 3.33 (with permission from Dassault Systems)

The Modify Tool:

The modify tool that appears when you create or edit a dimension (parameter) has several options. The options available are:

1. ▦ Spin the value up or down by a present amount.

2. ✓ Save the current value and exit the dialog box.

3. ✗ Restore the original value and exit the dialog box.

4. ▯ . Rebuild the model with the current value.

5. ±? Change the spin increment value.

6. ▣ Mark the dimension for drawing import.

    (with permission from Dassault Systems)

Note: While giving dimensions to the sketch keep into mind that you should always start with the smallest dimension first, and work your way to the largest.

Adding Angular Dimensions: Using the dimension tool, create angular dimension. Create a sketch. Select the lines where you have to enter the value of the angle. Modify tool dialog box appears.

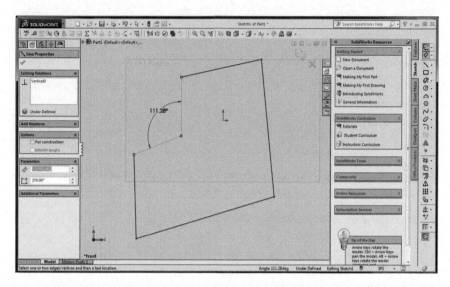

Fig. 3.34 (with permission from Dassault Systems)

Edit the angle to 80°.

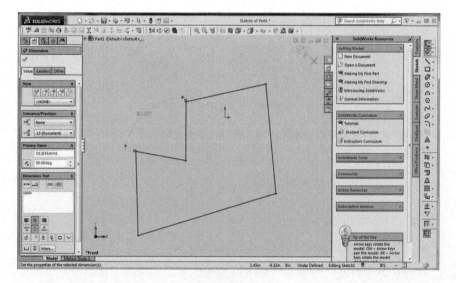

Fig. 3.35 (with permission from Dassault Systems)

Edit all the dimensions one by one by starting from the smallest.

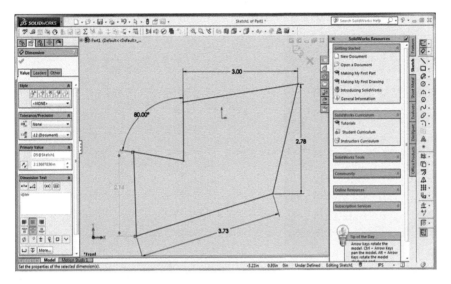

Fig. 3.36 (with permission from Dassault Systems)

Click to close the dialog box. The sketch is fully defined.

Chapter 4
Part Modeling Using Features

In this chapter, we will be covering the special features present in SOLIDWORKS after the Part Design using sketching. After successful completion of selection of plane and sketching, a part one can use the same for Extrusion, Revolve, Swept, Extruded cut, Revolved cut, etc. We will be dealing with all the features one by one in detail in this chapter with the help of figures.

4.1 Extruded Boss/Base

Extrudes a sketch or selected sketch contours in one or two directions to create a solid feature. Different end conditions can be seen while extrusion of the sketched part such as Blind, Up to Vertex, Up to surface, Offset from Surface, Up to Body, and Mid Plane.

Click on **New Document**.
Select the document **Part** and click **ok**.
Go to **Sketch** select the plane (**Front Plane**).
Right-click on the **Front Plane** and select **Normal To**.
Select **Corner Rectangle** from the Rectangular Section.
Create a **Corner Rectangle**.
Close the dialog box.
Go to the Features section and click on **Extruded Boss/Base**.

© Springer Nature Switzerland AG 2020
K. Kumar et al., *Mastering SolidWorks*, Management and Industrial Engineering,
https://doi.org/10.1007/978-3-030-38901-7_4

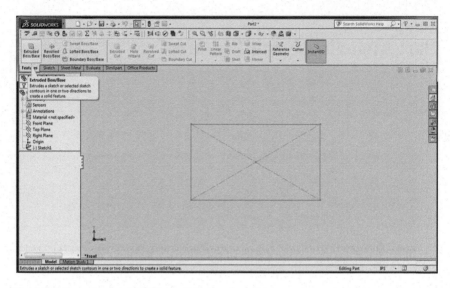

Fig. 4.1 (with permission from Dassault Systems)

Select a handle to modify parameters and by default end condition will be selected as **BLIND**.

Set depth as 0.60 in.

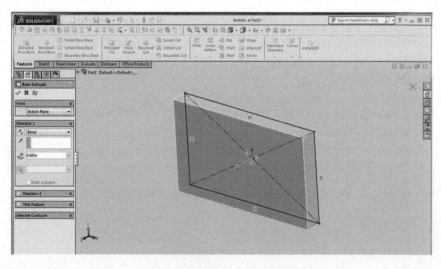

Fig. 4.2 (with permission from Dassault Systems)

Click the location in empty space to set the distance or select an entity to extrude up to it by clicking on the **Direction 2**.

Set the depth distance as 0.60 in and the user can see the difference between the two figures.

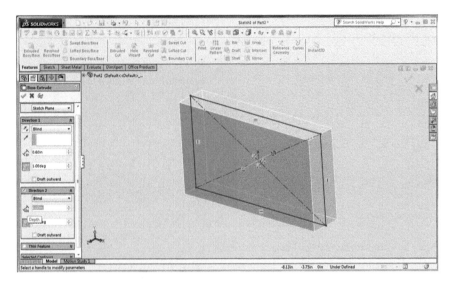

Fig. 4.3 (with permission from Dassault Systems)

Click on the end condition and change it from **Blind** to **Up to Vertex**.

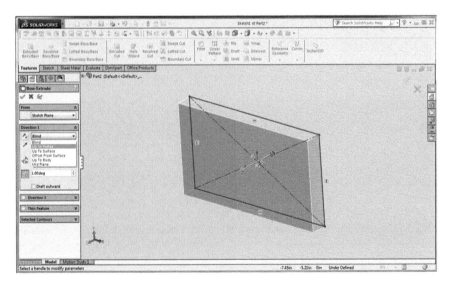

Fig. 4.4 (with permission from Dassault Systems)

Select a vertex to complete the specification of Direction 1.

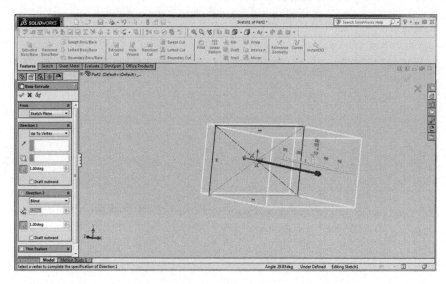

Fig. 4.5 (with permission from Dassault Systems)

Select a plane or surface to complete the specification of Direction 1.
Close the dialog box by clicking on the tick mark.

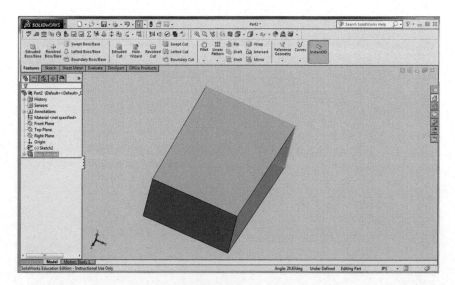

Fig. 4.6 (with permission from Dassault Systems)

Go to **feature manager design tree**.
Right-click on the **Extruded Boss/Base** and click on **Edit feature**.
Select the end condition in **Direction 1** as **Up to surface**.

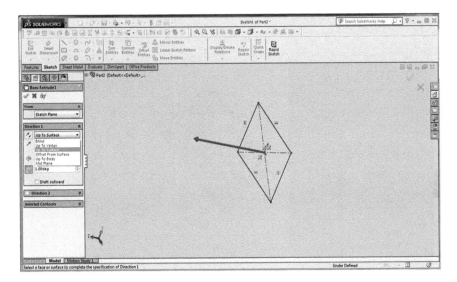

Fig. 4.7 (with permission from Dassault Systems)

Select a face or surface to complete the specification of Direction 1.
Close the dialog box by clicking on the tick mark.

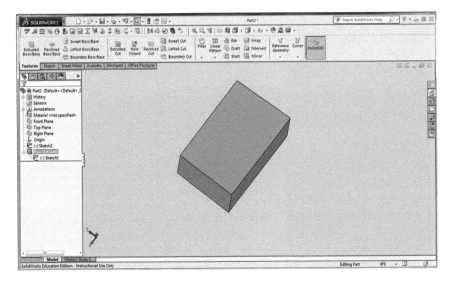

Fig. 4.8 (with permission from Dassault Systems)

Again go to the **feature design manager tree** and right-click on the
Boss-Extrude1.
Select **Edit feature**.

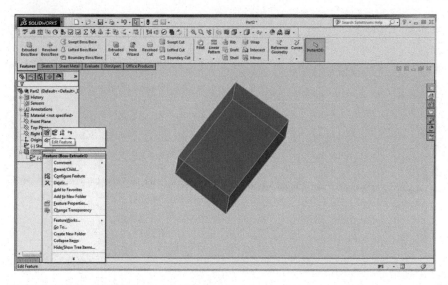

Fig. 4.9 (with permission from Dassault Systems)

Change the depth Direction 1 to **offset from surface**.

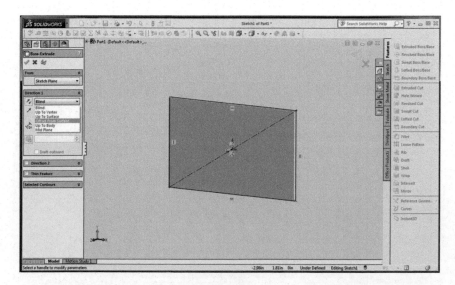

Fig. 4.10 (with permission from Dassault Systems)

Select a face or surface to complete the specification of Direction 1
Select the **front plane**.

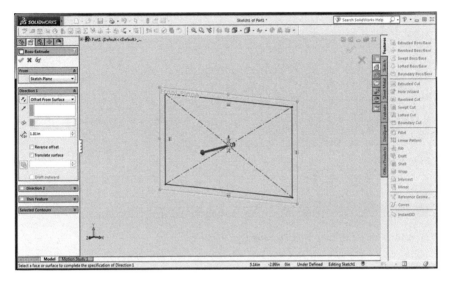

Fig. 4.11 (with permission from Dassault Systems)

Click **ok** to complete the end condition of **offset from surface**.

4.2   Revolve Boss/Base

This feature revolves a sketch or selected sketch contours around an axis to create a
solid feature. To start with Revolve Boss/Base feature follow the steps given below.

1. Go to Sketch.
2. Select a plane (Front Plane).
3. Click on Line command sketch figure with the help of line.
4. Double-click to end the Line command.
5. Edit sketch and click on centerline.
6. Create a centerline in front of the sketch drawn to revolve.

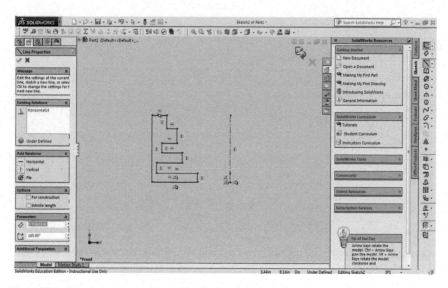

Fig. 4.12 (with permission from Dassault Systems)

7. Click ok to close the command and exit.
8. Go to the features section > Activate Revolve Boss/Base.

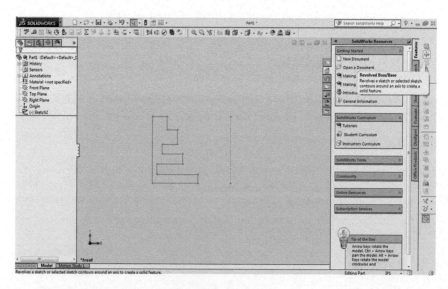

Fig. 4.13 (with permission from Dassault Systems)

9. Select an axis of revolution (centerline) about which sketch will revolve and set the parameters.

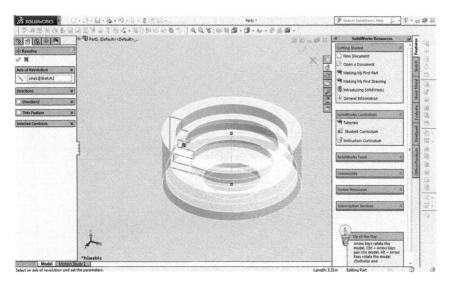

Fig. 4.14 (with permission from Dassault Systems)

10. Click tick mark to close and exit the revolve command.
11. Feature defined successfully.

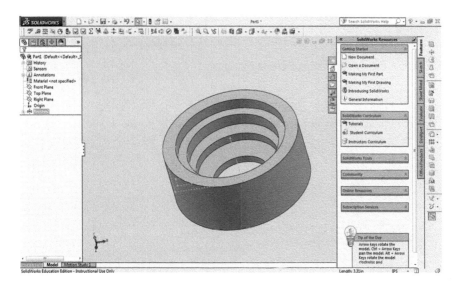

Fig. 4.15 (with permission from Dassault Systems)

4.3 Swept Boss/Base

This feature sweeps closed profile along an open or closed path to create a solid feature. This can be done by the following given steps one by one.

1. Select the plane as (Top Plane).
2. Go to sketch > Create a circle.
3. Close to end the command.
4. Create sketch_2 by selecting another plane as Front Plane.
5. Select Normal To.

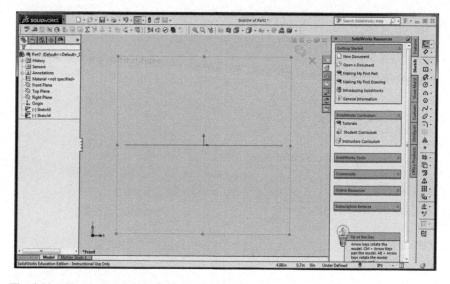

Fig. 4.16 (with permission from Dassault Systems)

6. Go to sketch > click on Spline.
7. Create a sketch by connecting the circle mid-point to the spline.

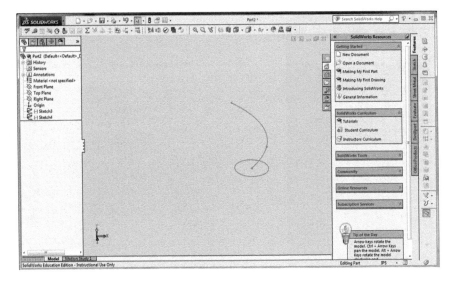

Fig. 4.17 (with permission from Dassault Systems)

8. Double-click to end the Spline.
9. Go to features > Select Swept Boss/Base.
10. Select sweep Profile (circle as Sketch3).
11. Select sweep path (spline as Sketch4).

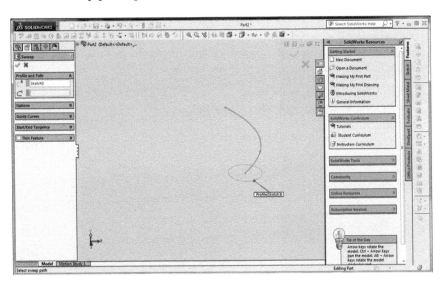

Fig. 4.18 (with permission from Dassault Systems)

12. Click on the tick mark and exit command.
13. Feature defined successfully.

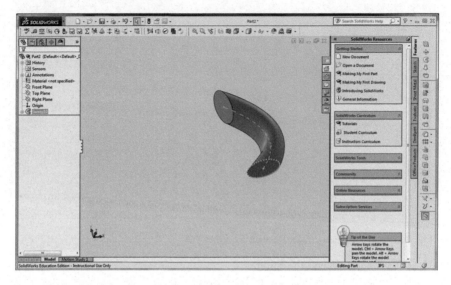

Fig. 4.19 (with permission from Dassault Systems)

4.4 Lofted Boss/Base

The feature adds material between two or more profile to create a solid feature. To use this feature, start with the steps given below.

1. Select the plane as (Front Plane).
2. Select the Reference geometry command.

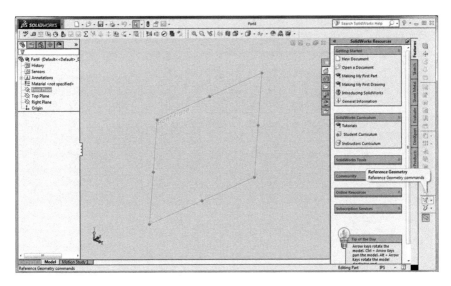

Fig. 4.20 (with permission from Dassault Systems)

3. Click on the front plane and drag it to some distance (3.60 inch).
4. Close the dialog box of the plane.

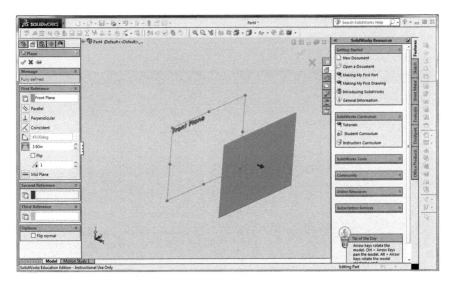

Fig. 4.21 (with permission from Dassault Systems)

5. Click on the plane 1 created and create a circle.

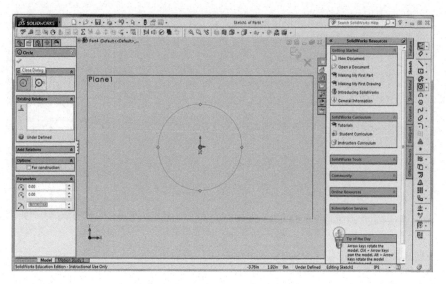

Fig. 4.22 (with permission from Dassault Systems)

6. Click ok to close and exit the dialog box.
7. Select the front plane from which reference is created and create a rectangle on the plane.

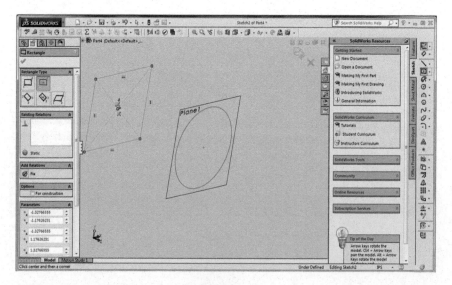

Fig. 4.23 (with permission from Dassault Systems)

8. Click ok to exit the rectangle command.
9. On both the plane (Front) and the reference plane, sketch is created.

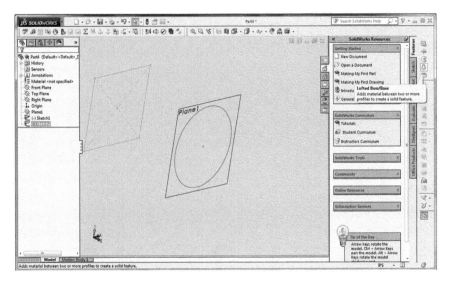

Fig. 4.24 (with permission from Dassault Systems)

10. Go to features section > Select the Loft command.
11. Select profiles as Sketch1 and Sketch2 to set attributes for loft.

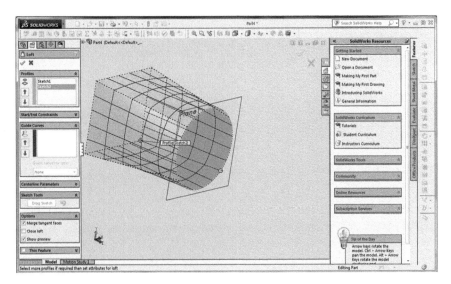

Fig. 4.25 (with permission from Dassault Systems)

12. Click ok.
13. Feature defined successfully.

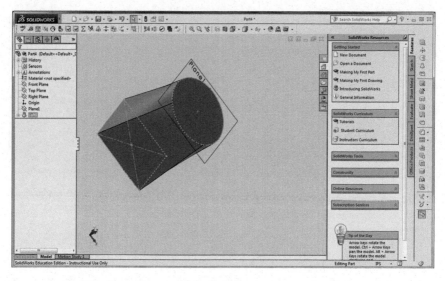

Fig. 4.26 (with permission from Dassault Systems)

4.5   Boundary Boss/Base

It adds material between profiles in two directions to create a solid feature. To start with the feature, follow the steps provided.

1. Select a plane (Top plane).
2. Create a Corner Rectangle on the plane.
3. Close the dialog box.

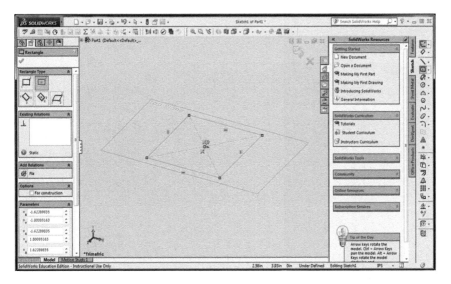

Fig. 4.27 (with permission from Dassault Systems)

4. Select another plane Front plane to create another sketch.
5. Click on spline to add splines points that shape the curve.

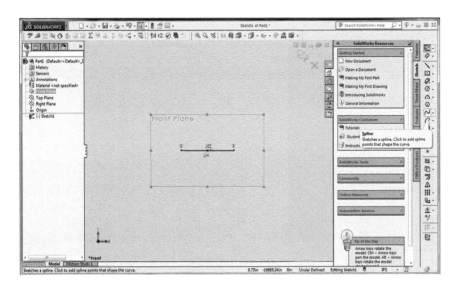

Fig. 4.28 (with permission from Dassault Systems)

6. Sketch 2 will be created with the spline.
7. Close the dialog box to exit.

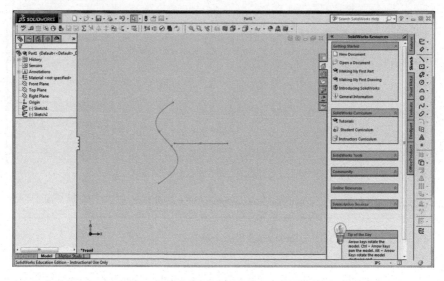

Fig. 4.29 (with permission from Dassault Systems)

8. Click on the Spline and the line of the sketch to pierce as shown in Fig. 4.30.

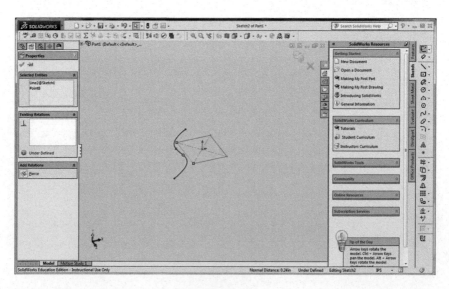

Fig. 4.30 (with permission from Dassault Systems)

9. Click ok after piercing.

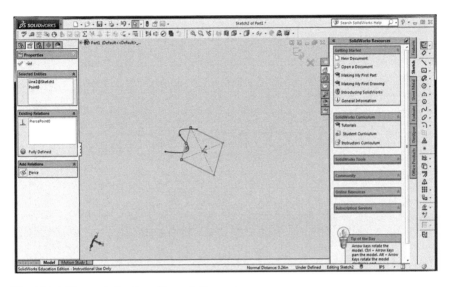

Fig. 4.31 (with permission from Dassault Systems)

10. Select third plane as Right plane to create a sketch.

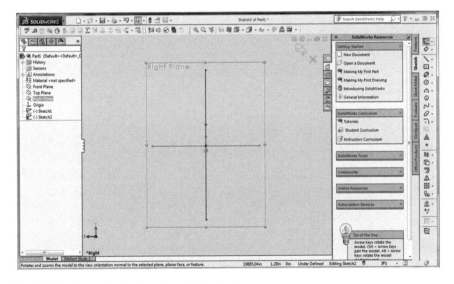

Fig. 4.32 (with permission from Dassault Systems)

11. Click on spline and create a spline on the plane.
12. Double-click to end the spline command.

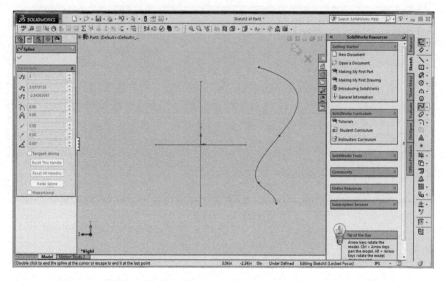

Fig. 4.33 (with permission from Dassault Systems)

13. Follow the same step and pierce the spline with the rectangle.

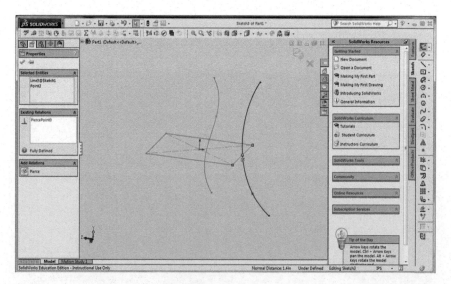

Fig. 4.34 (with permission from Dassault Systems)

14. Select Boundary Boss/Base from features section.
15. Select curve in Direction 1 as Sketch 1(Rectangle).
16. In Direction 2, select the Splines.

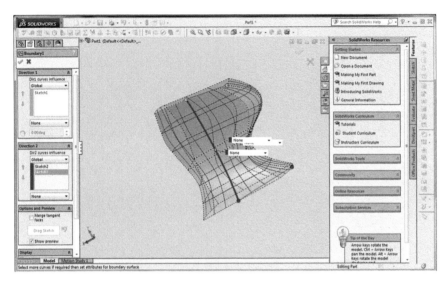

Fig. 4.35 (with permission from Dassault Systems)

17. Click ok to close the dialog box.
18. Boundary Boss/Base defined successfully.

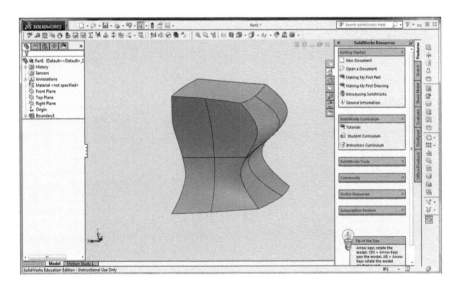

Fig. 4.36 (with permission from Dassault Systems)

Chapter 5
Sketch Entities

This chapter will consist of different types of entities in Sketch while designing a part or a model. Entities mainly include functions like removing of any unwanted sketch or a part from the drawn final model, modifying a part design after fully developing it as a final product, moving of the design and so on. Hence, for each there is a different command available in the command manager which we will be discussing here in details and show with the help of figures how it will work without any disarray by following the steps after each command.

There are mainly six entities available in SOLIDWORKS which include trim entities and extend entities, convert, offset, mirror, linear sketch pattern and move entities.

5.1 Trim Entities

This type of entity trims or extends sketch entity to be coincident to another, or deletes a sketch entity. There are several trimming options available in trim entities: power trim, corner trim, trim away inside, trim away outside and trim to closest.

To cut down the sketch or remove the part which is no more needed the designer can use trim entities. It can be found by selecting **Sketch > Trim entities**.

5.1.1 Power Trim

Power trim is used to remove portion of an entity that the user drags between intersections or to an endpoint. To use power trim.

1. Select a plane.
2. Create a sketch having number of horizontal and vertical lines with the use of Line command.

© Springer Nature Switzerland AG 2020
K. Kumar et al., *Mastering SolidWorks*, Management and Industrial Engineering,
https://doi.org/10.1007/978-3-030-38901-7_5

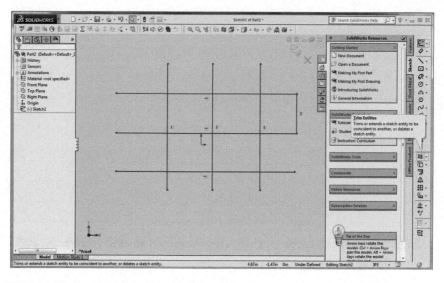

Fig. 5.1 (with permission from Dassault Systems)

3. Click on the Trim entities.
4. Select the entities to be removed by holding down cursor of your mouse.

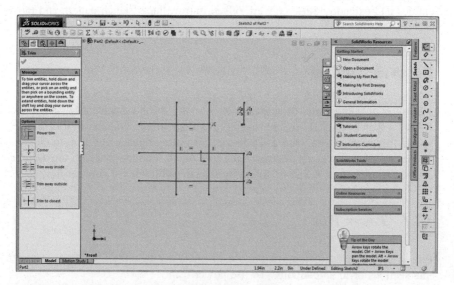

Fig. 5.2 (with permission from Dassault Systems)

5. More entities can be selected and removed by the same method.
6. Click ok to close the command.

5.1.2   Corner

This type of trim is used when corner is to be trimmed by selecting two entities. To use corner trim.

1. Select a plane and create a sketch with the use of Line command.
2. Select the corner from the trim entities.

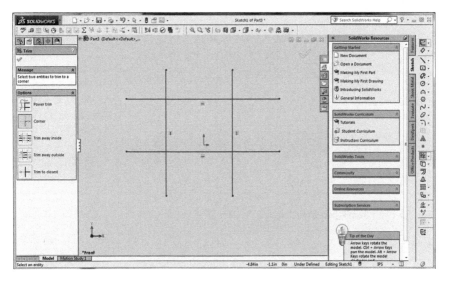

Fig. 5.3 (with permission from Dassault Systems)

3. Now select the two lines which are intersecting.

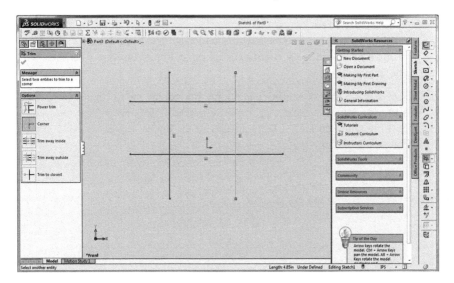

Fig. 5.4 (with permission from Dassault Systems)

4. After selecting one of the entities intersecting you can see a color change on the lines selected.

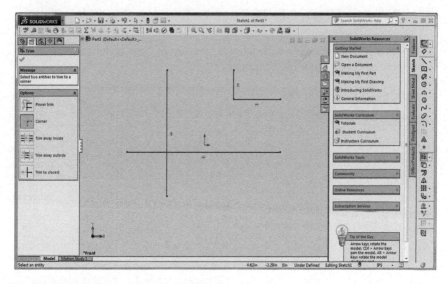

Fig. 5.5 (with permission from Dassault Systems)

5. Corner trim is done.
6. Click ok to complete and close the dialog box to exit.

5.1.3   Trim Away Inside

This option is used to remove the portions which are inside the boundaries. This can be performed by selecting two entities or a face and then selection of entities which are to be trimmed.

1. Select a plane and sketch lines including horizontal and vertical.
2. Go to trim entities > Select Trim Away Inside.
3. Select bounding entities. After selection of the entity, it will turn into sky blue color.

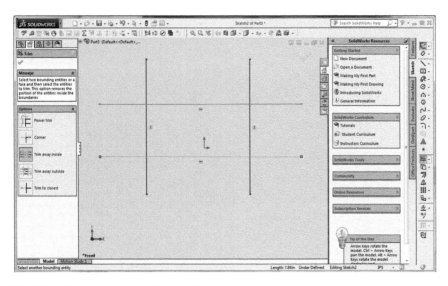

Fig. 5.6 (with permission from Dassault Systems)

4. Select or box-select the entity to be trimmed.
5. The entity selected will turn purple.

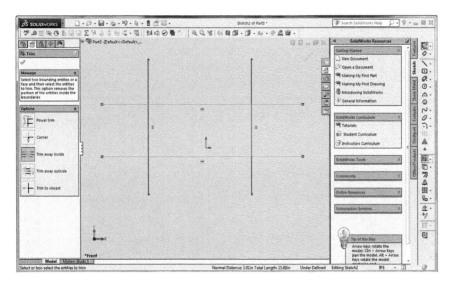

Fig. 5.7 (with permission from Dassault Systems)

6. After selection of both the bounding and entity to be trimmed, trim away inside
    can be seen.

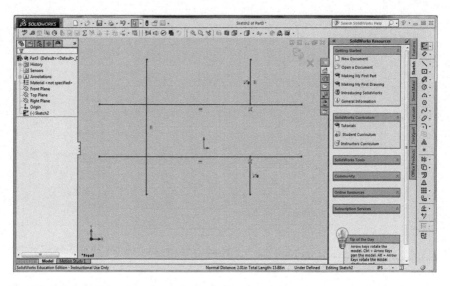

Fig. 5.8 (with permission from Dassault Systems)

5.1.4 Trim Away Outside

This option is used to remove the portion of the entities outside the boundaries. This can be done in the same way as trim away inside is performed.

1. Take the same sketch > select trim away outside.
2. Select the bounding entities (sky blue in color).

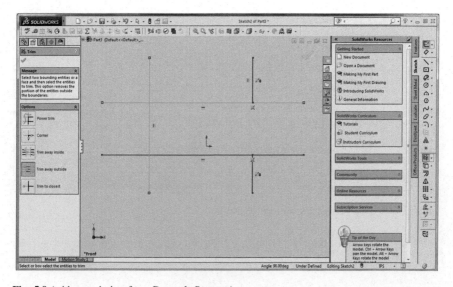

Fig. 5.9 (with permission from Dassault Systems)

Before selecting the entities to trim it should be kept in mind that the entity selected should intersect each boundary exactly once or not intersect either (to be deleted).

3. Select the entities to trim (shown in figure with red color).

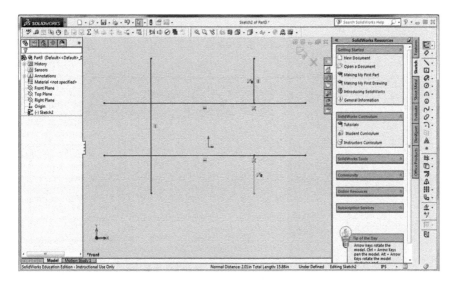

Fig. 5.10 (with permission from Dassault Systems)

4. After clicking on the trimmed entity it will be removed.
5. Click ok to complete the trim away outside command and Exit.

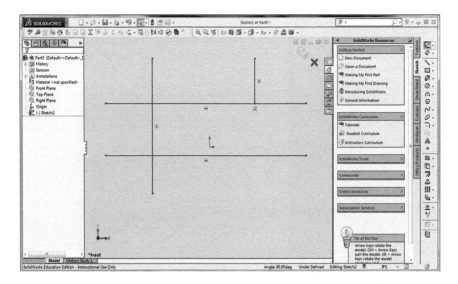

Fig. 5.11 (with permission from Dassault Systems)

5.1.5   *Trim to Closest*

This option is used to trim the portion of geometry between the boundaries. This is done by dragging the entity or selecting an entity which is to be trimmed to the nearest intersecting entity.

1. Create a sketch.
2. Sketch > Trim entities > Trim to closest.

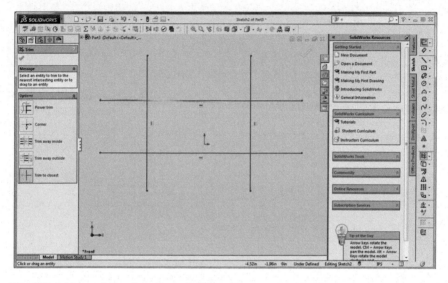

Fig. 5.12 (with permission from Dassault Systems)

3. Click on the entity to remove or drag the entity.
4. Select the entities one by one a scissor will appear beside the entity while selecting and selected lines turn purple.

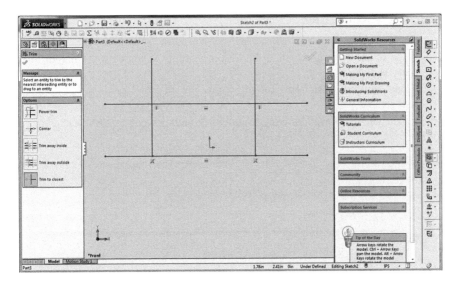

Fig. 5.13 (with permission from Dassault Systems)

5. Select all the unwanted entities to trim and create a rectangle.
6. Click ok to Exit.

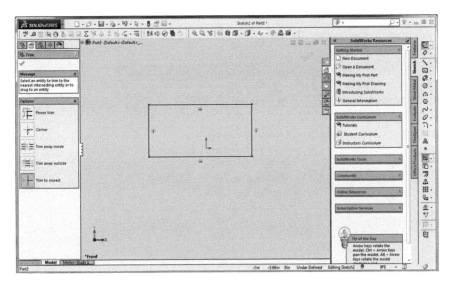

Fig. 5.14 (with permission from Dassault Systems)

5.2   Extend Entities

This type of entity is used when the designer wants to increase the length of the geometry sketched.

Extend entities can be found from the Tools menu > Select sketch Tools > Extend.

Or, can also be viewed by Sketch Toolbar > click Extend entities. It can be done in the following way given below.

1. Select a plane.
2. Create a Sketch.
3. Click on Sketch Toolbar and select Extend entities.

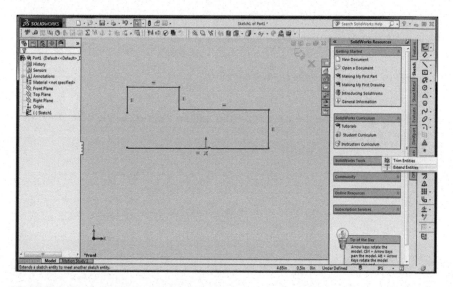

Fig. 5.15 (with permission from Dassault Systems)

4. Select the end which is nearest and can be extended.
5. Click to extend to the nearest intersection.

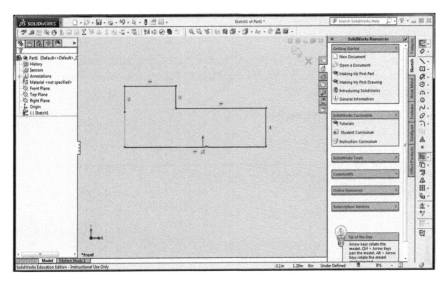

Fig. 5.16 (with permission from Dassault Systems)

6. Selected entity will turn orange in color and will be extended till the next intersection entity.
7. Follow the same step and extend as many lines as required to be extended.

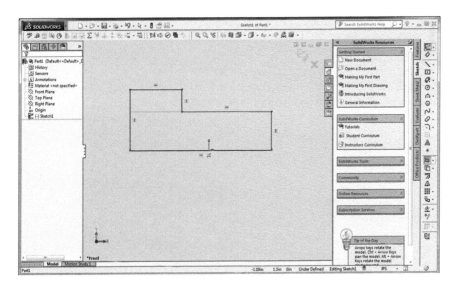

Fig. 5.17 (with permission from Dassault Systems)

5.3   Convert Entities

This type of entities converts selected model edges or sketch entities into sketch segments. To use this entity the designer has to follow the below given steps and proceed in the following way.

1. Select a plane and create a sketch.
2. Extrude the sketch.

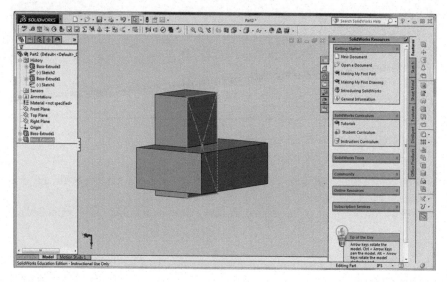

Fig. 5.18 (with permission from Dassault Systems)

3. Click on the convert entities command from the sketch tool.
4. Select the faces with the help of mouse button.

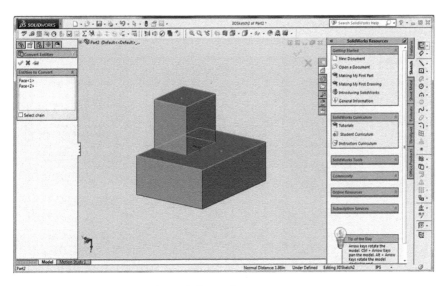

Fig. 5.19 (with permission from Dassault Systems)

5. The faces selected will be converted and can be seen in black color.
6. Select the edges to be converted and click ok.

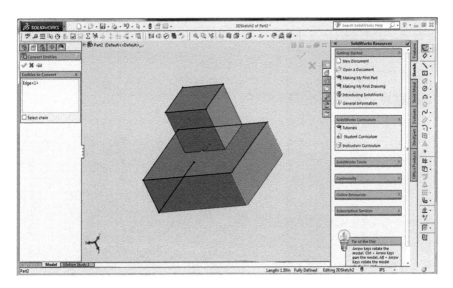

Fig. 5.20 (with permission from Dassault Systems)

7. Any number of faces, edges, or sketch entities can be converted with this entity.

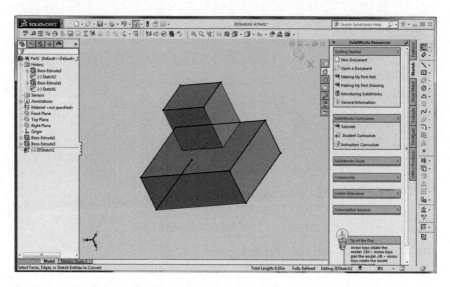

Fig. 5.21 (with permission from Dassault Systems)

5.3.1   Intersection Curve

It creates a sketch curve along the intersection of planes, solid bodies, and surface bodies. It generates intersection curve by selecting sketch entities.

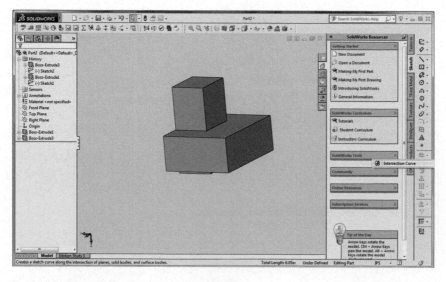

Fig. 5.22 (with permission from Dassault Systems)

5.4 Offset

It adds sketch entities by offsetting faces, edges, curves, or sketch entities a specifies distance. To start with offset follow the given below steps.

1. Select a plane.
2. Create a rectangle.

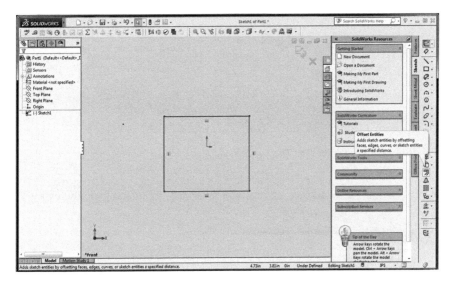

Fig. 5.23 (with permission from Dassault Systems)

3. Click on offset entities.
4. Edit the offset parameters. Click on all the sides to complete if it has been created with Line command.
5. Click and drag for dynamic control.

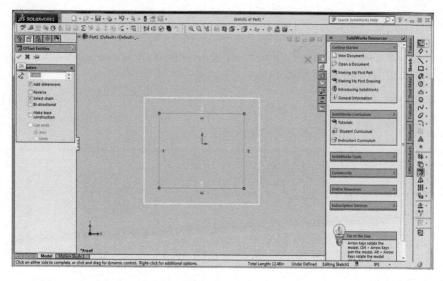

Fig. 5.24 (with permission from Dassault Systems)

6. Right-click for additional options and then click ok to complete the entity.

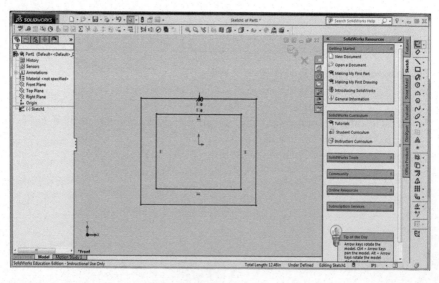

Fig. 5.25 (with permission from Dassault Systems)

5.5   Mirror Entities

This type of entity mirrors selected entities about a centerline. To start with mirror entity starts with below given procedure.

1. Select a plane.
2. Go to sketch > create a centerline > ok to exit or double-click.

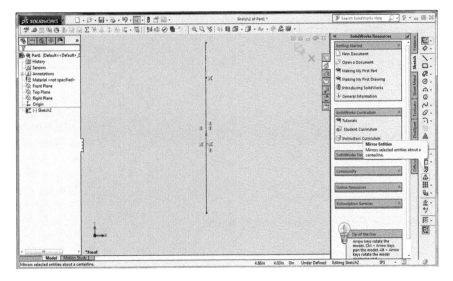

Fig. 5.26 (with permission from Dassault Systems)

3. Create a sketch as shown in Fig. 5.27.

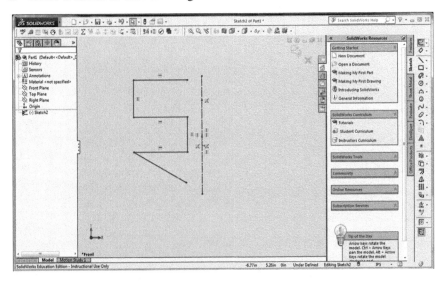

Fig. 5.27 (with permission from Dassault Systems)

4. Select mirror entities > click on entities to mirror.
5. Select the centerline to mirror about.

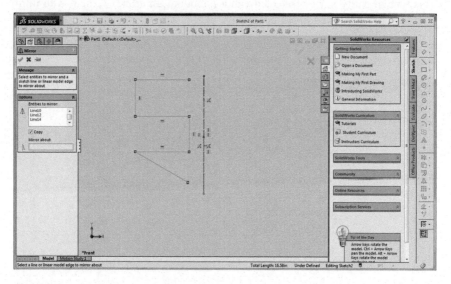

Fig. 5.28 (with permission from Dassault Systems)

6. Click on ok to create mirror entity.

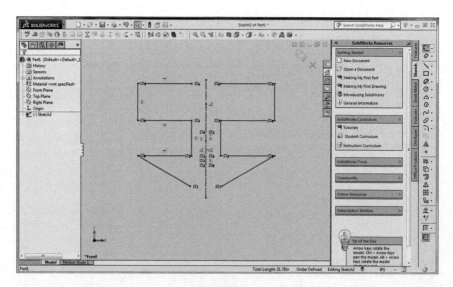

Fig. 5.29 (with permission from Dassault Systems)

5.6 Linear Sketch Pattern

Adds a linear pattern of sketch entities. Linear pattern can be created by following the given below steps carefully.

1. Select a plane.
2. Go to sketch > create a sketch with one horizontal line and one vertical line.
3. Double-click to exit the Line command.
4. Go to sketch > click on circle and create a circle on vertical line as shown in Fig. 5.30.

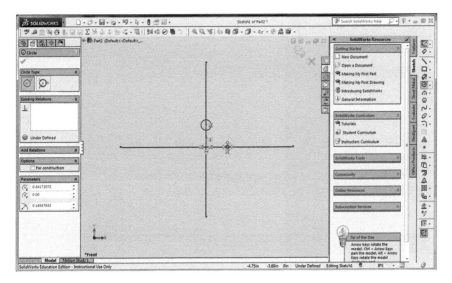

Fig. 5.30 (with permission from Dassault Systems)

5. Click ok to exit the command.
6. Select linear sketch pattern same circle will be drawn with same dimension in the given axis.
7. Reverse direction can be chosen to move the circle to the other side and edit spacing between the circles.

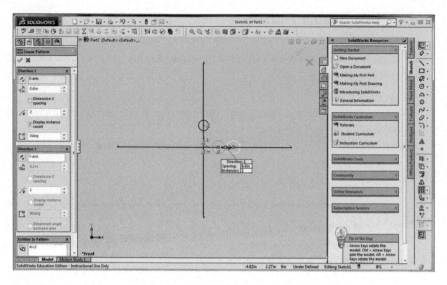

Fig. 5.31 (with permission from Dassault Systems)

8. Click on circular sketch pattern which adds circular spacing on the sketch entities.
9. Edit the number of instances and spacing as required.
10. Circular pattern will be created as can be seen from Fig. 5.32.

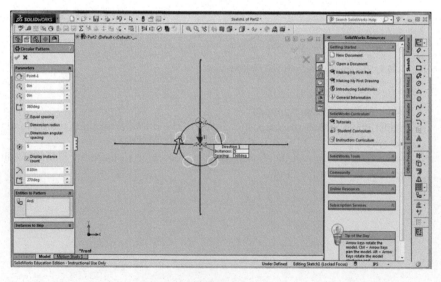

Fig. 5.32 (with permission from Dassault Systems)

5.7 Move Entities

It moves sketch entities and annotations. To create move entity create a sketch by selecting a plane.

1. Create a horizontal and a vertical line intersecting each other.
2. Click on Move entities.

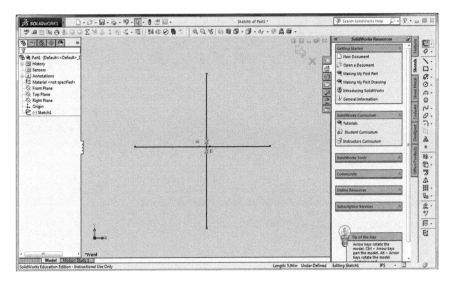

Fig. 5.33 (with permission from Dassault Systems)

3. Select the entity which is to be moved.
4. Select the starting point and sketch items or annotations.

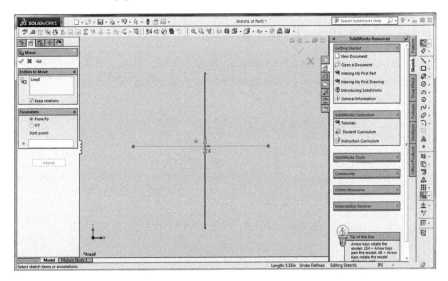

Fig. 5.34 (with permission from Dassault Systems)

5. Click to define the destination of the move.
6. The user will be able to move the entity upward downward as desired.

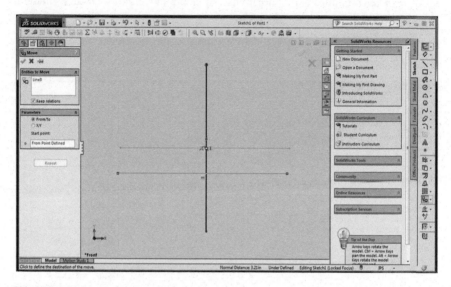

Fig. 5.35 (with permission from Dassault Systems)

5.8 Copy Entities

This entity helps in copying sketch entities and annotations. If same sketch is to be designed, then one can use copy entity that will give the same sketch and as per the dimensions given. To start with copy entities follow the steps given below.

1. Select a plane.
2. Go to sketch > create a circle.
3. Select Copy entities.
4. Click on the entity which is to be copied.
5. Set the parameters, i.e., click to define the destination of the copy.

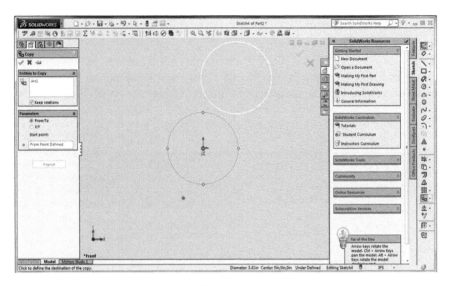

Fig. 5.36 (with permission from Dassault Systems)

6. Click ok to complete the entity.

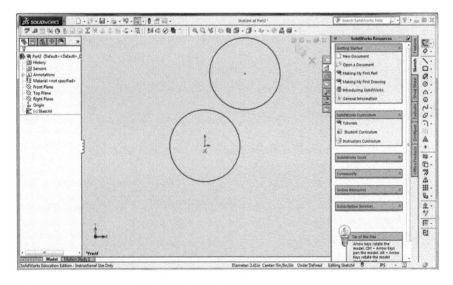

Fig. 5.37 (with permission from Dassault Systems)

5.9   Rotate Entities

This type of entities helps in rotating sketch entities and annotations. Take the same sketch and click on rotate entity. Select the entities to rotate and set the parameters such as angle in degrees and center of rotation at which the sketch is to be rotated. Click on the tick mark to complete the entity.

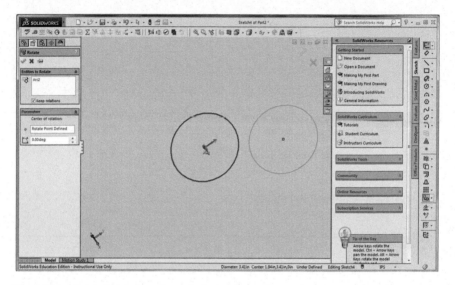

Fig. 5.38 (with permission from Dassault Systems)

5.10   Scale Entities

It sketches scale entities and annotations. To start with scaling designer has to proceed in the following way.

1. Select the plane.
2. Create a circle > use copy entity to create another circle with same dimensions.
3. Click on scale entity.
4. Select entity to be scaled.
5. Click on entity about which it is to be scaled.
6. Increase the scaling parameter. It can be seen that with the increment of scaling the size of the circle increases.

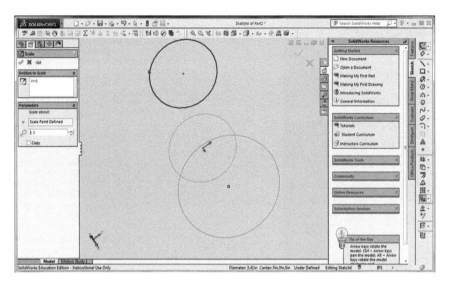

Fig. 5.39 (with permission from Dassault Systems)

5.11   Stretch Entities

It stretches sketch entities and annotations. With the help of stretching entity, it is seen that the sketch selected will stretch about the base defined.

1. Take the same sketch used for scaling and click on the Stretch entity.
2. After selecting the entity to be stretched click on the parameter from which it is to be stretched.
3. Centerline can be seen in Fig. 5.40 while stretching about the based selected.
4. Click to define the destination of the move.

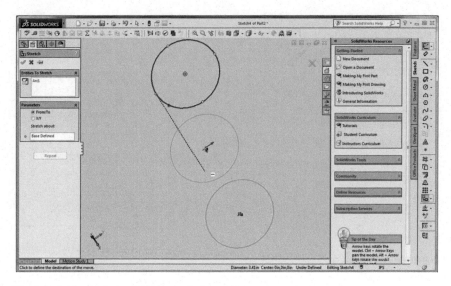

Fig. 5.40 (with permission from Dassault Systems)

5. Click ok to complete the feature successfully.

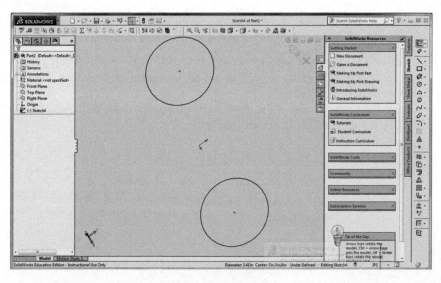

Fig. 5.41 (with permission from Dassault Systems)

Chapter 6
Features Including Cuts

At the end of this chapter, one will be able to know briefly about different features using cuts like Extruded Cut, Hole Wizard, Revolved Cut, Swept Cut, Lofted Cut, and Boundary Cut. Cutting or creating a hole in the sketch model is done with the help of the features stated. Once the two main Boss features like Extruded Boss/ Base and Revolved Boss/Base gets completed creation of a cut is done to represent the removal of material. Cut features are crated in the same way as boss/base features get created.

6.1 Extruded Cut

The menu for creating a cut feature by extruding is identical to that of creating a boss. The only difference is that a cut removes material while a boss adds it. Other than that distinction, the commands are the same, and this cut represents a slot.

Extruded Cut can be found from:

1. From the **insert** menu, select **Cut**, **Extrude**.

© Springer Nature Switzerland AG 2020   141
K. Kumar et al., *Mastering SolidWorks*, Management and Industrial Engineering,
https://doi.org/10.1007/978-3-030-38901-7_6

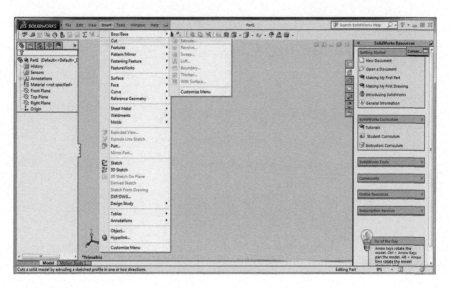

Fig. 6.1 (with permission from Dassault Systems)

2. Or, on the **features** toolbar, choose **Extrude cut** .

To start with Extruded cut, follow the given steps carefully.

1. Select a plane (Front plane).
2. Click on the plane section and click on Normal To.
3. Go to sketch toolbar, click on **Corner Rectangle**.
4. Create a **corner rectangle**.
5. Now click on **features toolbar** and select **Extruded Boss/Base**.
6. Modify the depth dimension to 0.60 inch.
7. Click on the tick mark [image] to close the dialog box.

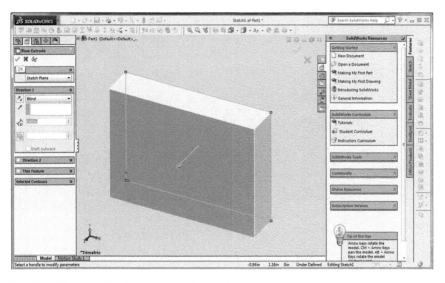

Fig. 6.2 (with permission from Dassault Systems)

8. After clicking on, the [✓] user can see that in the features section toolbar, Extruded Cut option gets activated.
9. Click on the **Extruded cut** and activate the command.
10. Now select a plane, a planar face, or an edge on which to sketch the feature cross section or the user can also select an existing sketch to use for the feature.
11. Once selection of plane is done go to sketch toolbar and create a circle on the respective plane.

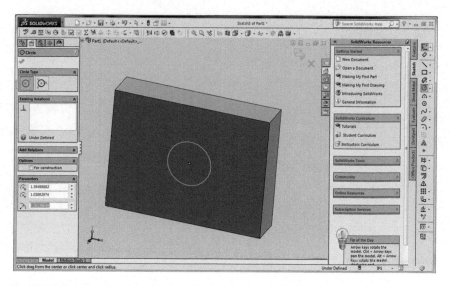

Fig. 6.3 (with permission from Dassault Systems)

12. Close the dialog box by clicking on the [checkmark icon] .
13. Click on **features toolbar** and select **Extruded Cut**.

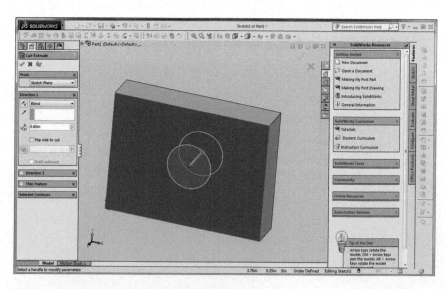

Fig. 6.4 (with permission from Dassault Systems)

14. Modification of depth can be done as per the dimension. By default, BLIND will be the end condition but it can be changed as Through All, Up to Next, Up to Vertex, Up to surface, Up to Body, Mid Plane.
15. Click to close the dialog box and view the final sketch design.

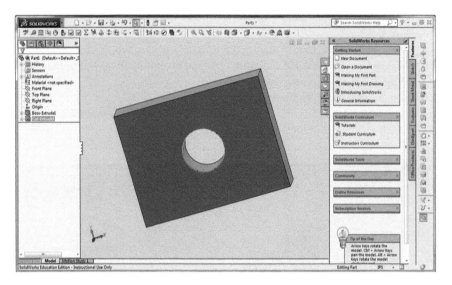

Fig. 6.5 (with permission from Dassault Systems)

6.2 Hole Wizard

The Hole Wizard is used to create specialized holes in a solid. It can create simple, tapered, counter bored and countersunk holes using a step-by-step procedure. In this example, the Hole Wizard will be used to create a standard hole.

The Hole wizard creates shaped holes, such as counter bore and countersunk types. The process creates two sketches. One defines the shape of the hole, and other, a point, locates the center.

The Hole wizard requires a face to be selected or pre-selected, not a sketch. It can be found from:

1. From the **insert** menu choose **features**, **Hole Wizard**…

2. Or choose the **Hole Wizard** [icon] tool on the f**eatures toolbar**.

To start with the command, follow the steps given below one by one and start creating a hole wizard.

1. First step is to select a plane. Select plane as **FRONT** Plane.
2. Go to **sketch toolbar** and select **corner rectangle**.
3. Click **ok** to close the dialog box.
4. Then click on the **features toolbar** to select the **Extruded Boss/Base**.
5. Edit the **depth distance** from 0.10 to 0.30 inch.
6. Click **ok** and close the dialog box.
7. After creating **Extruded Boss/Base** in the features section hole wizard icon gets activated.
8. Click on Hole Wizard to start creating a hole of any type and dimension.
9. When Hole wizard will be clicked, type and position will appear to the left side of the graphics window screen.
10. In the **Type,** enter **counter bore**.
11. Shift to the **positions** tab to position the hole. Use the dimensions and other sketch tools to position the hole or slot.

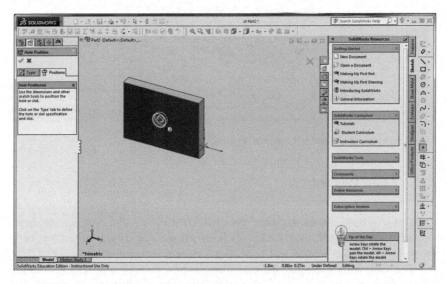

Fig. 6.6 (with permission from Dassault Systems)

12. Click on the **Type tab** to define the hole or slot specification and size.
13. Select size as ¼.
14. Move the cursor down and type **end condition** as **Through All**.
15. Select **standard** as **ANSI Inch**.
16. Click ok and close the dialog box.
17. In the left-hand side of the feature manager design tree, it will be seen clearly that the feature design is completed for CBORE for ¼ Binding Head Machine Screw 1.

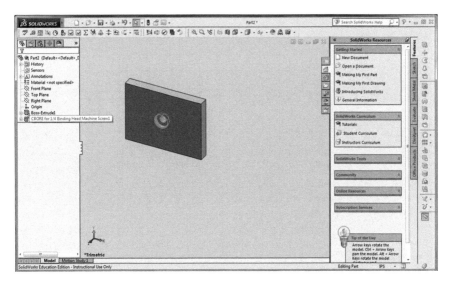

Fig. 6.7 (with permission from Dassault Systems)

6.3 Revolved Cut

It cuts a solid model by revolving a sketched profile around an axis. It can be found from:

1. From the **insert** menu choose **features**, **Revolved Cut**.
2. Or choose the **Revolved Cut** 🔄 from the **features toolbar**.

To start with revolved cut, follow the given steps carefully.

1. Select **Front plane** and make it **normal to** the screen.
2. Go to **sketch toolbar** and click on the **line command**.
3. With the help of line command, create 2 horizontal and 1 vertical line as if you are drawing a rectangle but it's not get completed because of less number of vertical lines than horizontal.
4. Double-click on the command or right-click and select to end the line command.
5. Now click on **3 tangent arc** and with the help of the arc create a sketch as shown in Fig. 6.8.

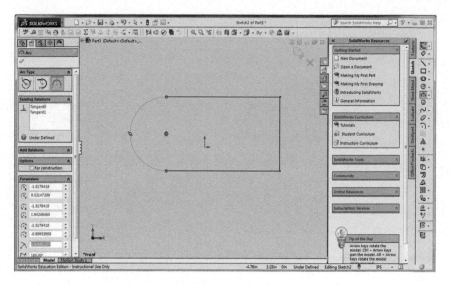

Fig. 6.8 (with permission from Dassault Systems)

6. Click **ok** to close the dialog box.
7. Click on **centerline** and create a centerline above the sketch and at the left side of the sketch.

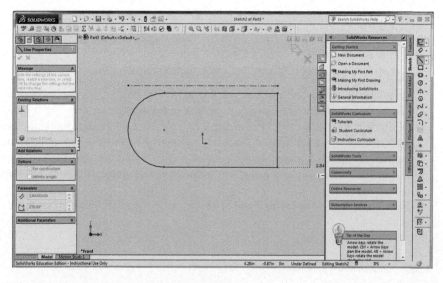

Fig. 6.9 (with permission from Dassault Systems)

8. Make the centerlines equal to the sketch created and click ok and end the command.
9. Go to features toolbar and select **Revolved Boss/Base**.

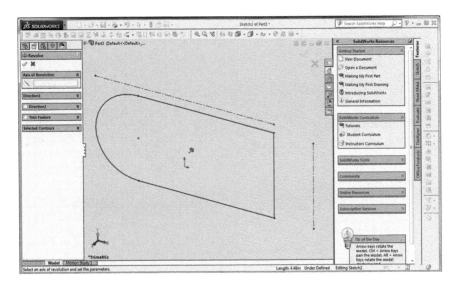

Fig. 6.10 (with permission from Dassault Systems)

10. Select the **Axis of revolution** as **Line 5**.

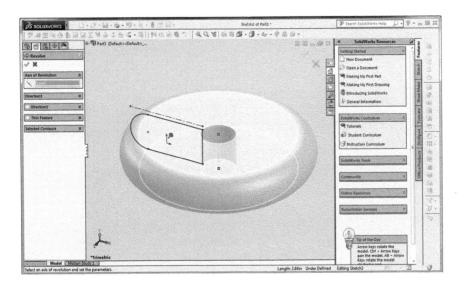

Fig. 6.11 (with permission from Dassault Systems)

11. Selection of **Line 4** as **Axis of revolution** will give the below result as shown in Fig. 6.12.

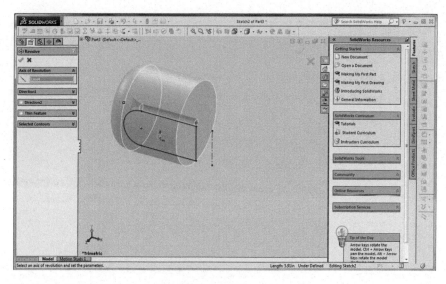

Fig. 6.12 (with permission from Dassault Systems)

12. Difference in the design can be seen from both the above diagrams when axis of revolution is changed. Go back to the same axis of revolution as line 5 and continue the sketching process.

13. Click on the arrow given [⟳] to change the **direction** of the **depth** and edit the angle in degrees to 280.

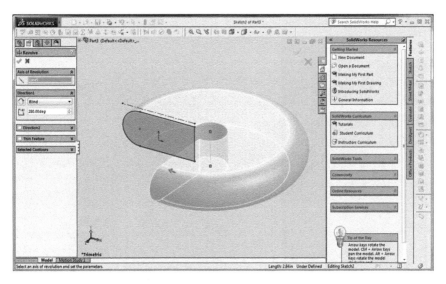

Fig. 6.13 (with permission from Dassault Systems)

14. Click **ok** to close the dialog box.
15. Select a plane, after selection it will turn blue.

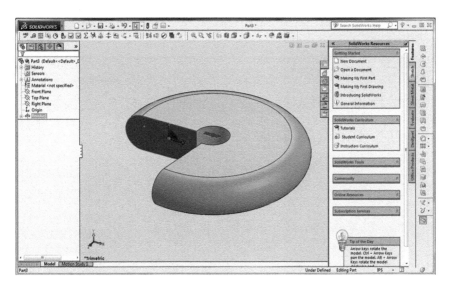

Fig. 6.14 (with permission from Dassault Systems)

16. Go to **sketch** and create a **circle** on the plane selected.

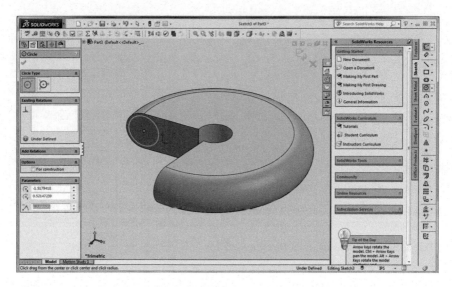

Fig. 6.15 (with permission from Dassault Systems)

17. Click on the icons shown below to view the **temporary axes** .

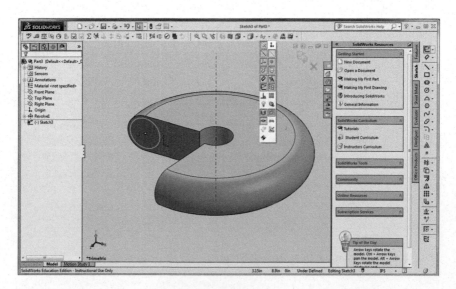

Fig. 6.16 (with permission from Dassault Systems)

18. Go to **features toolbar** and click on **Cut Revolve**. Select the **temporary axes** as **Axis of Revolution**.

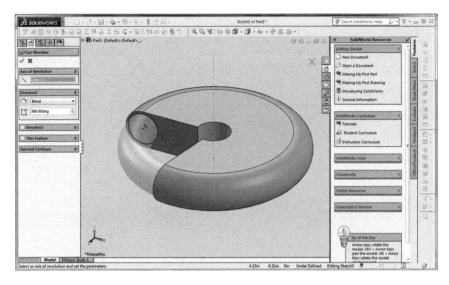

Fig. 6.17 (with permission from Dassault Systems)

19. Change the **depth direction** to see the difference in the movement of the circle and edit the angle.

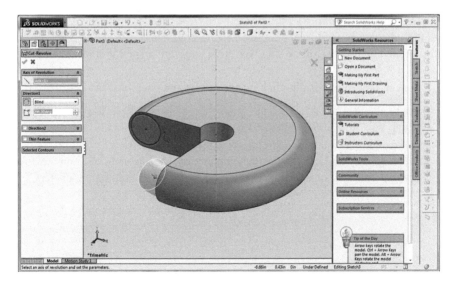

Fig. 6.18 (with permission from Dassault Systems)

20. Click **ok** and close the dialog box. Rotate and zoom to see the completely
    defined revolved cut.

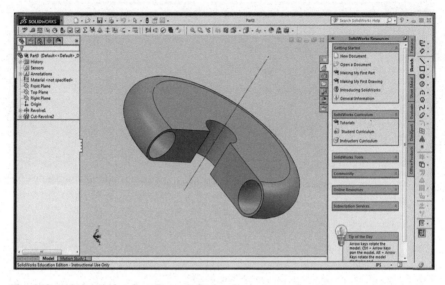

Fig. 6.19 (with permission from Dassault Systems)

6.4 Swept Cut

Swept cut helps in cutting a solid model by sweeping a closed profile along an open
or closed path. To start swept cut, we need to consider two planes one by one and
then sketch onto it. So follow the given steps and start the part design for swept cut.

1. Select **Top plane**. Right-click on the cursor and click on **Edit feature** .

2. Make the plane **Normal to** the screen and create a **corner Rectangle**.
3. Click **ok** and close the dialog box.
4. Go to **features toolbar** and click on **Extruded Boss/Base**.
5. In the **Direction 1** change the default direction. To change it, click on the down
   arrow; many options will appear to modify the sketch and its parameters.

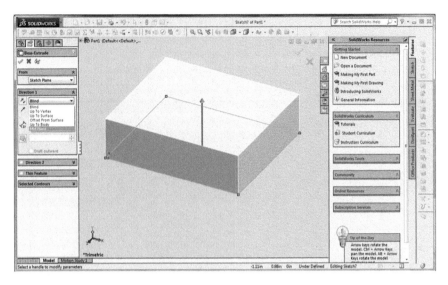

Fig. 6.20 (with permission from Dassault Systems)

6. Select **Mid Plane** and edit the depth (D1) to 1.00 in.
7. Click **ok** to close the dialog box.

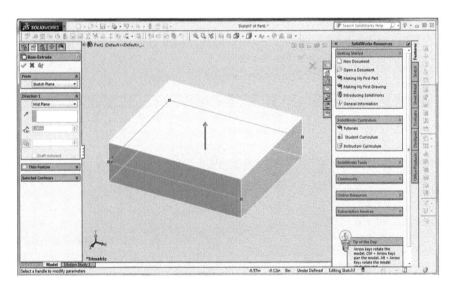

Fig. 6.21 (with permission from Dassault Systems)

8. Rotate the sketch and zoom in or zoom out to see the sketch from all the sides.
9. Click on the **Top Plane** and Edit feature by making it **Normal To** the screen.
10. Go to **sketch** section and click on **Line** command.
11. Start from the point and sketch in the same way as shown in given Fig. 6.22.

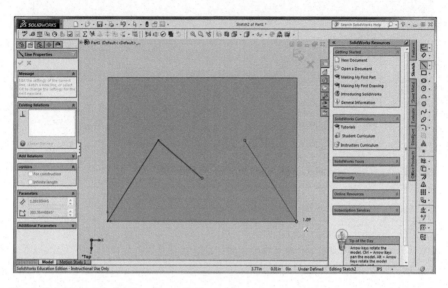

Fig. 6.22 (with permission from Dassault Systems)

12. Click **ok** to close the dialog box. **Sketch2** created as can be seen in the left side of the feature manager design tree.

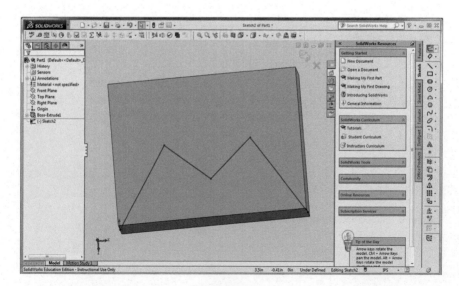

Fig. 6.23 (with permission from Dassault Systems)

13. Go to sketch and select **Sketch Fillet** and select entities to fillet. Modify the parameters if needed. Click **ok** and close the dialog box after completing the fillet part.

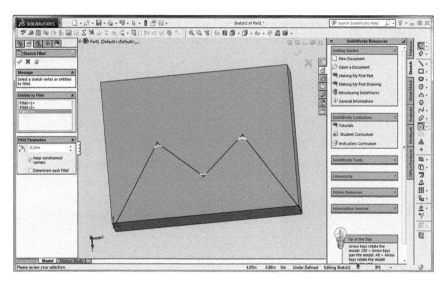

Fig. 6.24 (with permission from Dassault Systems)

14. Now go to **Front Plane** and select **Normal To**.

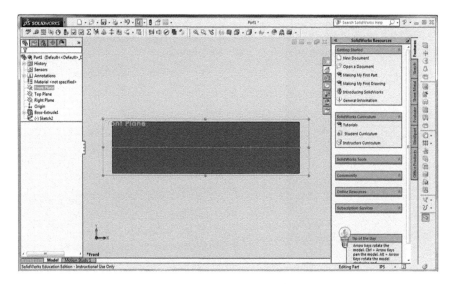

Fig. 6.25 (with permission from Dassault Systems)

15. Create another sketch by choosing **circle** command.
16. Create a circle in the mid of the point from where the sketch 2 was created.

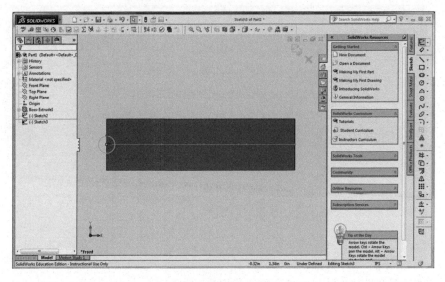

Fig. 6.26 (with permission from Dassault Systems)

17. Rotate and zoom the part after completing all the sketches on the graphics window screen after exiting the sketch. If the designer will not exit the sketch, it will be difficult for him to create swept cut because it will not be activated unless and until user exits the sketch.

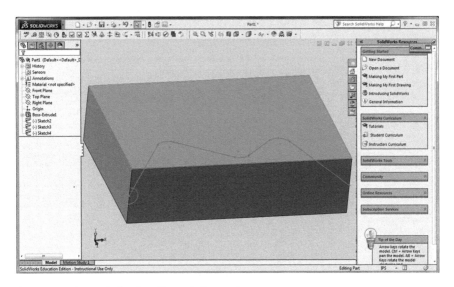

Fig. 6.27 (with permission from Dassault Systems)

18. Go to **features toolbar** and select **Swept cut**. It will ask to select **profile** and **path** for creation of the same.
19. Click on the circle (**sketch 4**) as profile and to select the **path** open the Part 1 by clicking on ⊟◌ **Part1 (Default<<Default>_** and the user will be able to see all the sketches created by him. So, select the lines created by clicking on **sketch 2** as whole.

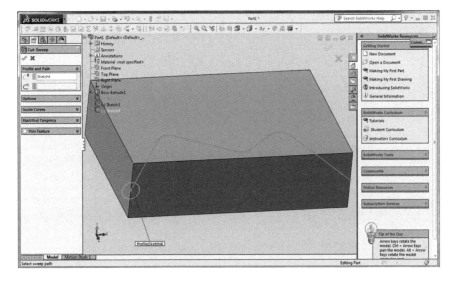

Fig. 6.28 (with permission from Dassault Systems)

20. Then move to **Thin feature** and click onto it.

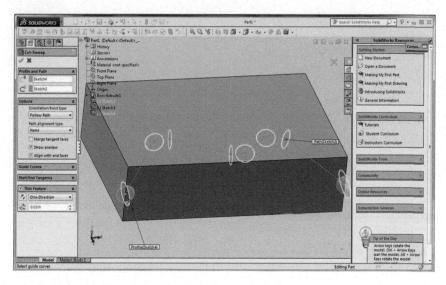

Fig. 6.29 (with permission from Dassault Systems)

21. Click **ok** to close the dialog box and view the **swept cut** created.

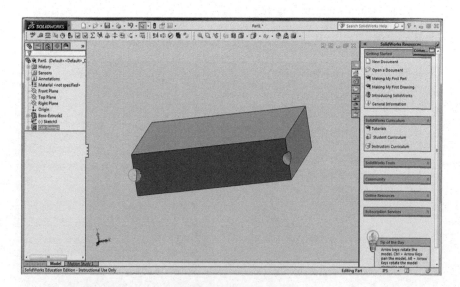

Fig. 6.30 (with permission from Dassault Systems)

22. Click on the **display style** and move down to **hidden lines visible**. Hidden lines visible will help the user look to the cut performed by displaying all the edges of the model.

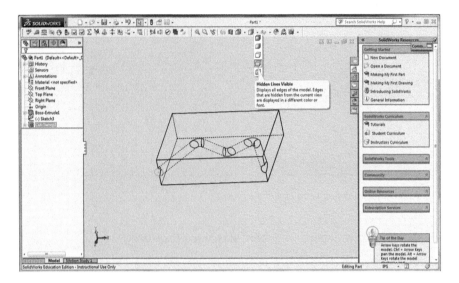

Fig. 6.31 (with permission from Dassault Systems)

6.5   Lofted Cut

Lofted cut is a feature which cuts the solid model by removing material between two or more profiles.

It can be found from:

1. From the **insert** menu choose **features**, **Lofted Cut**.
2. Or choose from the **features toolbar**.

To start lofted cut:

1. Select the plane (**Front Plane**) and make it **Normal to** the screen.
2. Go to **sketch toolbar** and select **corner rectangle**.
3. Create a **corner rectangle** and give dimensions with the help of **smart dimensions**.
4. Unit of the dimensions can be changed by the user as per the requirement.

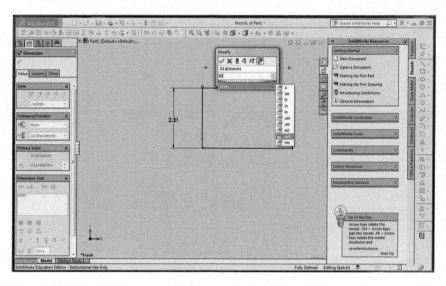

Fig. 6.32 (with permission from Dassault Systems)

5. Go to **features toolbar** and select **Extruded Boss/Base**.

6. Change the **direction** 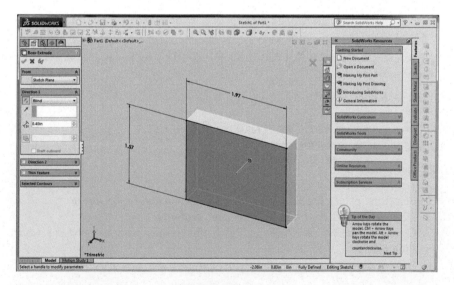 and edit the **depth**. Click **ok** to close the dialog box.

Fig. 6.33 (with permission from Dassault Systems)

7. Click on the **face** to create a sketch.

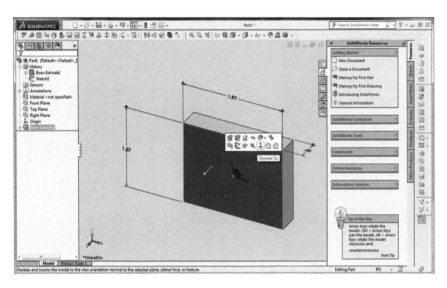

Fig. 6.34 (with permission from Dassault Systems)

8. Go to **sketch toolbar**. Select **line** command and create a sketch by touching line with the edges.

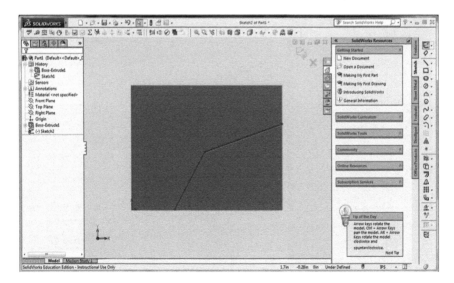

Fig. 6.35 (with permission from Dassault Systems)

9. With the help of **smart dimension** give dimension to each and every line with
   respect to the edge distance. Click **ok**.

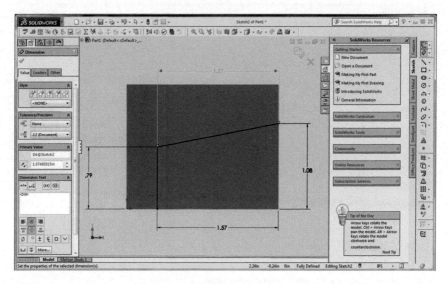

Fig. 6.36 (with permission from Dassault Systems)

10. Now select the **Top Plane** and click **Normal To** the screen.

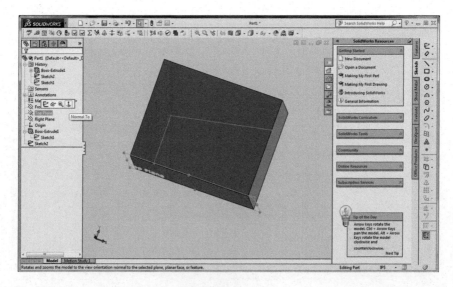

Fig. 6.37 (with permission from Dassault Systems)

11. Go to sketch and select Point ✳ .

12. Create appoint in the middle of the edge. Existing relation will appear as **Midpoint** in the tab on left side of the screen. Click **ok**.

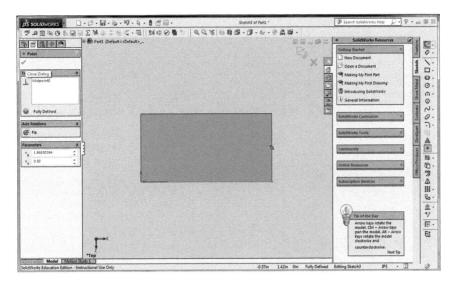

Fig. 6.38 (with permission from Dassault Systems)

13. Select **profile** to create Cut-Loft so by pressing CTRL click on the lines (**open Group <1>**) and then click on the **point**.

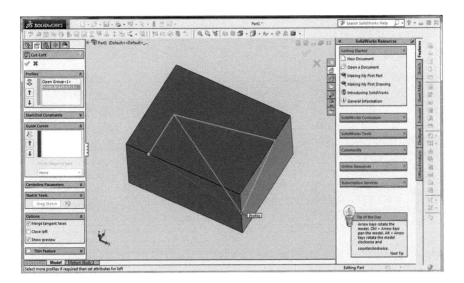

Fig. 6.39 (with permission from Dassault Systems)

14. When the user will click ok [✓] a tab will appear on the screen asking which bodies to keep. Select **Selected** bodies.

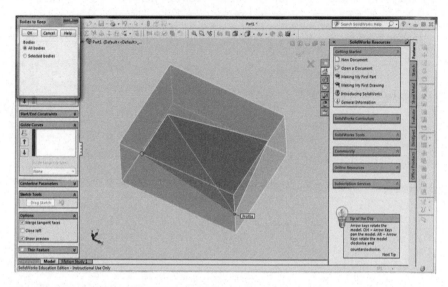

Fig. 6.40 (with permission from Dassault Systems)

15. Then it will ask whether Body 1 or Body 2. Select **Body 1**.

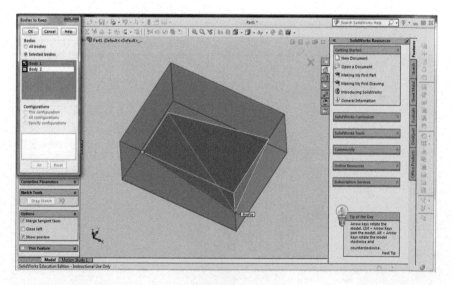

Fig. 6.41 (with permission from Dassault Systems)

16. Then click **ok,** lofted cut will be created successfully.

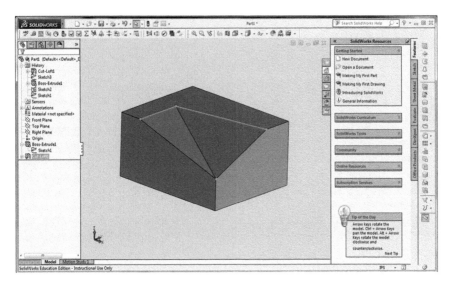

Fig. 6.42 (with permission from Dassault Systems)

6.6   Boundary Cut

It cuts the solid model by removing material between profiles in two directions. It can be found from:

1. From the **insert** menu choose **features**, **Boundary Cut**.
2. Or choose from the **features toolbar**.

To start boundary cut:

1. Select plane as (**Right Plane**) and click **normal to** the screen.
2. Go to **sketch toolbar** and select **centerline** and make it for **infinite construction**.
3. Then after sketching a centerline create a sketch as shown in the below figure with the help of **line** command.

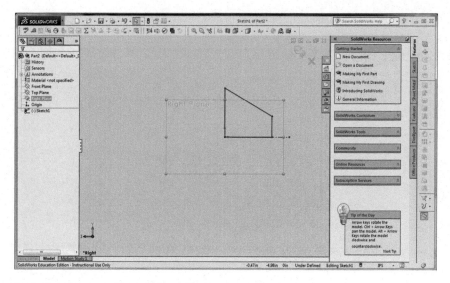

Fig. 6.43 (with permission from Dassault Systems)

4. Click on **features toolbar** and select **Revolve Boss/Base**. **Exit sketch**.

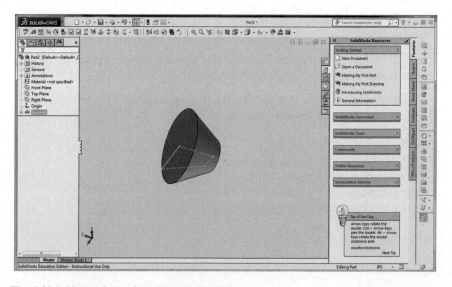

Fig. 6.44 (with permission from Dassault Systems)

5. From the **features toolbar** select **Boundary cut**. At least two profiles must be selected for creating a boundary cut. So first select the upper edge and then select the lower edge.

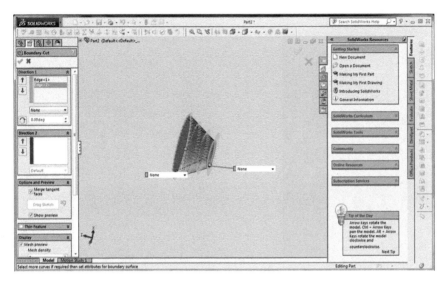

Fig. 6.45 (with permission from Dassault Systems)

6. Click ok to close the dialog box.
7. Boundary cut defined successfully.

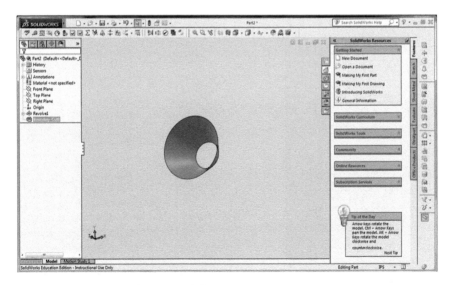

Fig. 6.46 (with permission from Dassault Systems)

Chapter 7
Thin Features

In this chapter, we will be dealing with features including Fillet, Pattern, Rib, Shell, Draft, Intersect and Wrap. After completion of part design, some more features have been included to do the required modifications in design with the help of above-said features. We will be discussing each with an example to make it clear how it works and how the user can create different part designs.

7.1 Fillet

It helps in creating rounded internal or external face along one or more edges in solid or surface feature. It can be done by selecting one or more than one edge of the sketch defined. Fillet is found from:

1. From the **insert** menu, select **Feature**, **Fillet**.
2. Or on the **Features** toolbar, choose **Fillet**.

To start with Fillet, follow the below-given steps.

1. Select a **Plane** (Front Plane).
2. Go to **sketch**.
3. Select **line** and create a sketch.
4. Go to **features toolbar** and click on **Extruded Boss/Base**.
5. Edit the **depth** and click **ok** to close the sketch.
6. Then again go to **features toolbar** and select **Fillet**.
7. Select the edges of the sketch to fillet and edit the **fillet parameter** as per the dimensions provided.

© Springer Nature Switzerland AG 2020
K. Kumar et al., *Mastering SolidWorks*, Management and Industrial Engineering,
https://doi.org/10.1007/978-3-030-38901-7_7

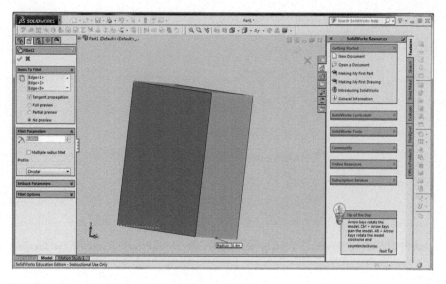

Fig. 7.1 (with permission from Dassault Systems)

8. Click **ok** to close the dialog box.
9. Fillet will be created on the edges selected as can be seen from Fig. 7.2.

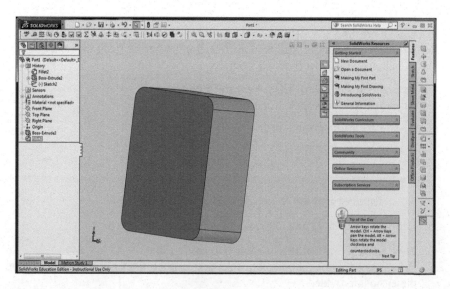

Fig. 7.2 (with permission from Dassault Systems)

7.2   Chamfer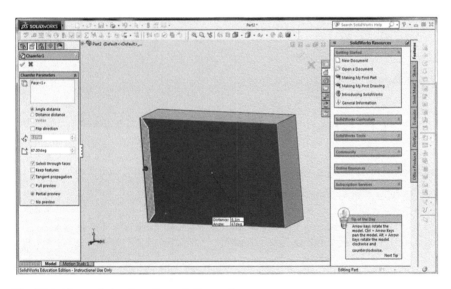

Chamfer creates a bevel feature along an edge, chain of tangent edges, or a vertex.
It can be found from:

1. From the **insert** menu, select **Feature**, **Chamfer**.
2. Or on the **features toolbar**, select **Chamfer** by dropping down the arrow button
   of Fillet.

   Chamfer can be created by

1. Select a **plane** (Front Plane).
2. Create a **corner rectangle**.
3. Go to **features toolbar**, select **Extruded Boss/Base**. Edit the **depth**   .
4. Select a face or an edge to chamfer. A face   is selected **Face<1>**. The user
   can select one or more than parameters to chamfer.
5. Edit the **depth**   and the **angle**   .

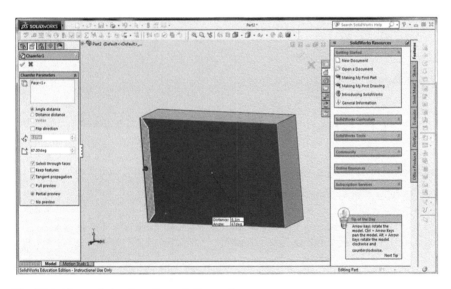

Fig. 7.3 (with permission from Dassault Systems)

6. Click **ok** 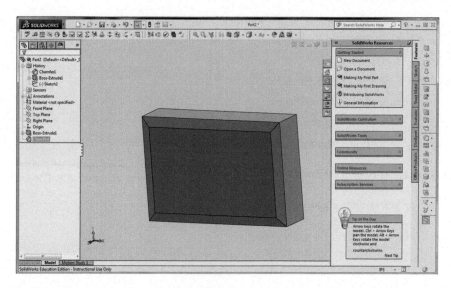 to close the dialog box.

7. Chamfer is created and features defined successfully.

Fig. 7.4 (with permission from Dassault Systems)

7.3   Draft

Draft tapers model faces by a specified angle, using a neutral plane or a parting line. To create draft, a neutral plane is needed and a face is to be selected to create a draft feature successfully.

To start with draft given below steps needed to be taken into consideration.

1. Select a **plane** (Front Plane).
2. Create a **sketch** with the help of **Line** command.
3. **Extrude** it by editing the **depth** and click **ok** to close the dialog box.
4. Go to **features toolbar**, select **Draft**.
5. Select the **neutral plane** as **Face<1>** and **Face<2>** to **draft** .
6. **Neutral plane** can be seen **pink** in color whereas, **faces to draft** are seen in **blue** color.
7. After selection of both the faces, click on **Apply**.

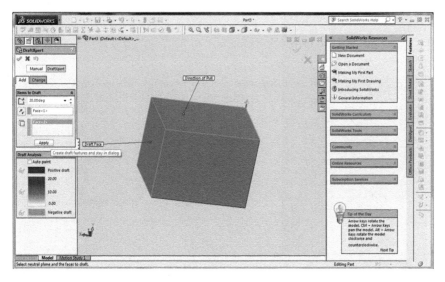

Fig. 7.5 (with permission from Dassault Systems)

8. At angle of **20**°, draft will be created.

9. Click on **Draft** from the **feature manager tree** and select **Edit Feature** .

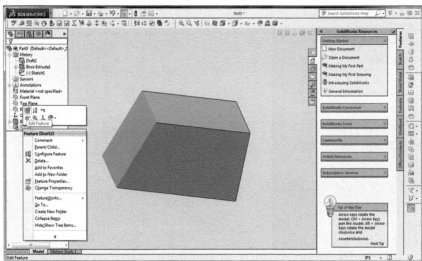

Fig. 7.6 (with permission from Dassault Systems)

10. Edit the angle and make it as 30°. Select more number of faces to draft.

11. Click **ok** ✅ to close the dialog box.

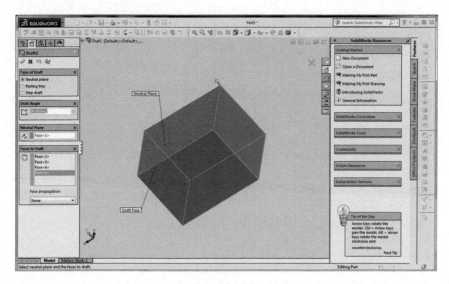

Fig. 7.7 (with permission from Dassault Systems)

12. **Draft** is created successfully.

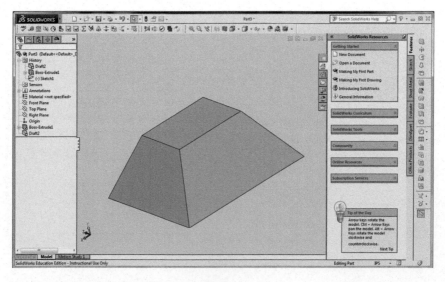

Fig. 7.8 (with permission from Dassault Systems)

7.4 Rib

It adds thin-walled support to a solid body. It can be found from the **features toolbar**, select **Rib**.

1. Select a **Plane** (**Front Plane**).
2. Go to sketch with the help of **Line** command and create a **sketch**.

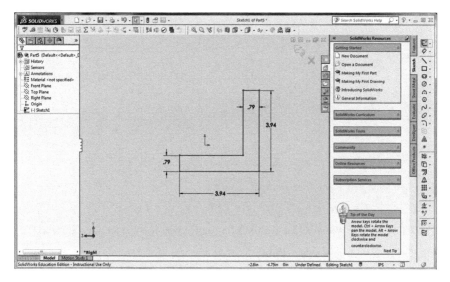

Fig. 7.9 (with permission from Dassault Systems)

3. After creating the sketch, click on **smart dimensions**.
4. Edit the dimensions.
5. Go to features toolbar, select **Extrude Boss/Base**.
6. Edit the **depth** in both the directions, i.e., Direction 1 and Direction 2 equally.
7. Or select the handle to modify parameters.

8. Click **ok** 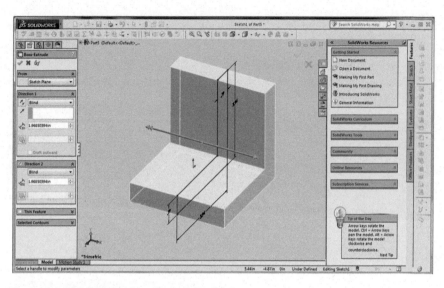 to close the dialog box.

Fig. 7.10 (with permission from Dassault Systems)

9. Click on the plane selected in starting and view it by selecting **Normal To**.
10. Go to sketch and select **Point** 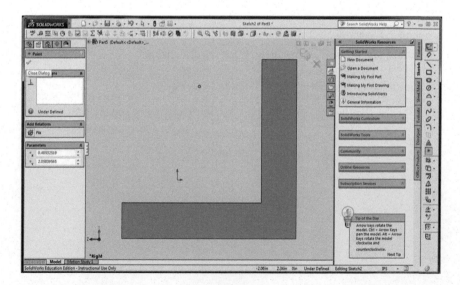.

Fig. 7.11 (with permission from Dassault Systems)

11. Edit the dimension width as 0.79 inches and height as 3.84 inches.
12. Click ok to close dialog box.

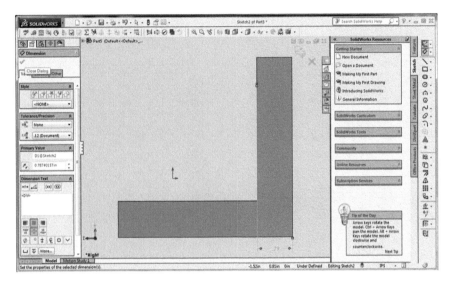

Fig. 7.12 (with permission from Dassault Systems)

13. Select the **base face** and click on the **point**.
14. Click on the point again it should be in the mid of the edge.

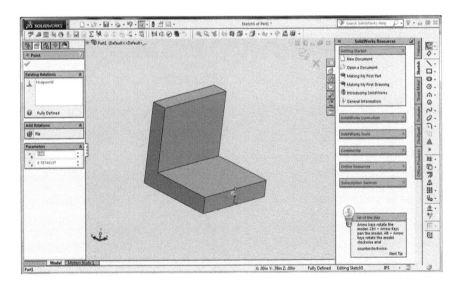

Fig. 7.13 (with permission from Dassault Systems)

15. Click **ok** 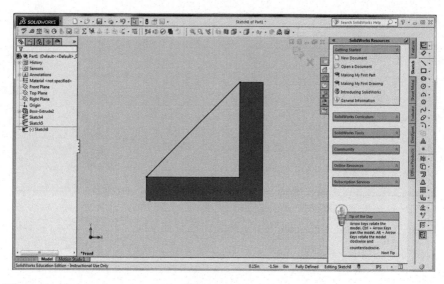 and then select Line to connect both the points.

Fig. 7.14 (with permission from Dassault Systems)

16. Set **direction**, **thickness,** and other parameters to create the rib.

17. **Flip the direction** and edit the thickness of the rib. Thickness can be provided either in one direction, in second direction or both the directions. Here both the sides are selected for thickness .

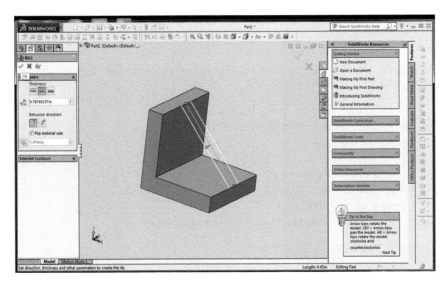

Fig. 7.15 (with permission from Dassault Systems)

18. Click ok .

19. **Rib** feature defined successfully.

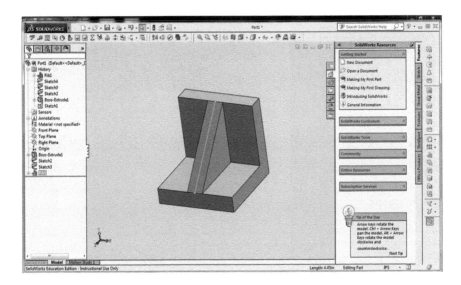

Fig. 7.16 (with permission from Dassault Systems)

7.5 Shell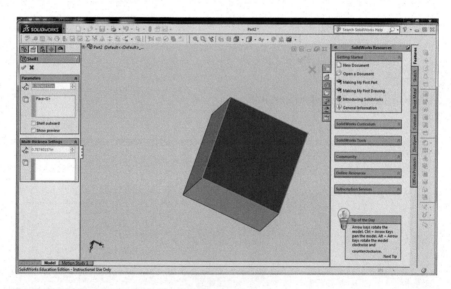

It removes material from a solid body to create a thin-walled structure. It can be found from:

1. From the **insert** menu, Select **features, Shell**.
2. Or on the **features toolbar**, select **Shell**.

   To start with Shell.

1. Select a **plane** (Front Plane).
2. Go to sketch. Select corner rectangle.
3. Create a **corner Rectangle**.
4. Go to **features toolbar**, select **Extruded Boss/Base**.

5. Select the **face** to create shell.

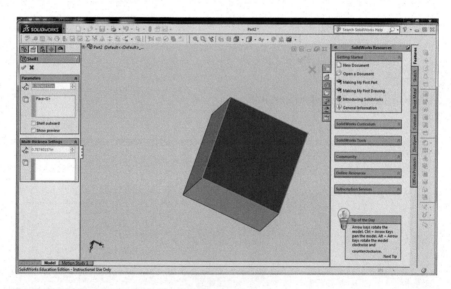

Fig. 7.17 (with permission from Dassault Systems)

6. Click **ok** ✅ .

7. **Shell** defined successfully.

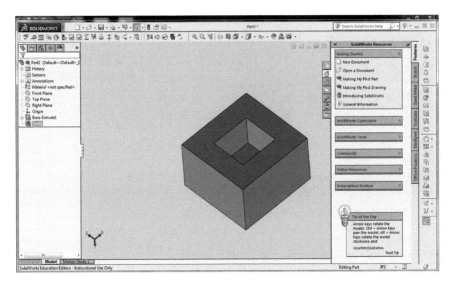

Fig. 7.18 (with permission from Dassault Systems)

7.6   Wrap 🗄

Wrap helps in sketching closed contours onto a face. It can be found from:

1. Go to **insert** menu, Select **Features**, **Wrap**.
2. Or on the **features toolbar**, select **Wrap**.

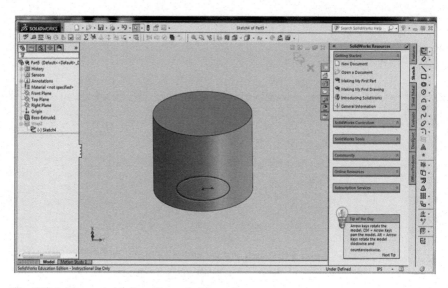

Fig. 7.19 (with permission from Dassault Systems)

3. Go to sketch, select **circle** and create a **circle**.
4. Extrude it with **Extruded Boss/Base** command.
5. Go to **features toolbar**, select **Wrap**.
6. Select another plane with respect to which wrap is to be created.
7. Click **ok**.
8. **Wrap** created.

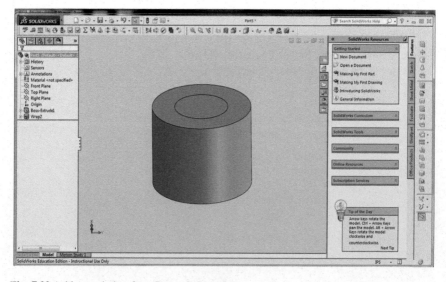

Fig. 7.20 (with permission from Dassault Systems)

7.7   Intersect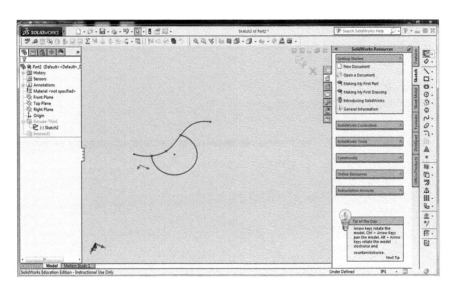

It intersects surfaces, planes and solids to create volume. Intersect can be found from:

1. Go to **insert** menu, select **feature**s, **Intersect**.
2. Or on the **features toolbar**, select **intersect**.

    To create intersect feature in part design.

1. Select a **plane**.
2. Go to sketch, select spline and create a **spline** on the selected plane.
3. Click ok to close the dialog box. Select arc and create a **3 tangent arc**.

4. Click **ok** .

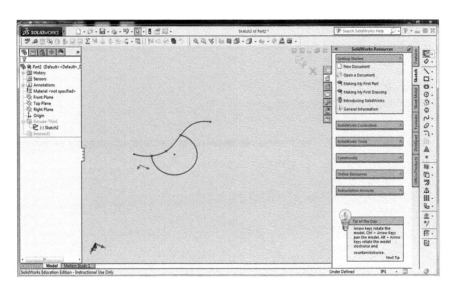

Fig. 7.21 (with permission from Dassault Systems)

5. Go to features toolbar, **Extrude** the sketch.
6. Click ok. Select Intersect to create intersection between the spline and the 3-tangent arc.
7. Select both the sketch for intersection.
8. Click on **Intersect**.

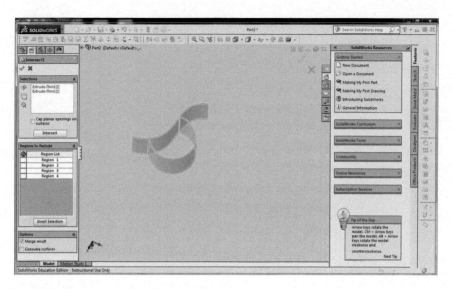

Fig. 7.22 (with permission from Dassault Systems)

9. Hit **ok**.
10. Intersect defined successfully.

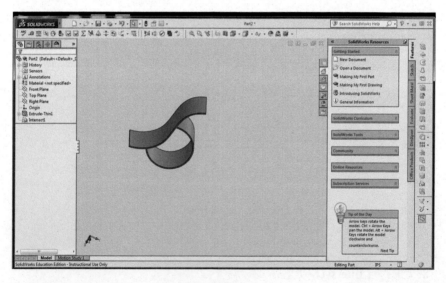

Fig. 7.23 (with permission from Dassault Systems)

Chapter 8
Introduction to Sheet Metal

Sheet Metal tool allows the designer to quickly create sheet metal part designs using simple design process and helps on saving time and development costs and increases productivity.

In this chapter, we will be discussing about Sheet Metal features like base flange, conversion of part into sheet metal, edge flange, Corners, Miter Flange, Extruded Cut and simple Hole. SOLIDWORKS models and automatically input bend radius, thickness and K-factor. It designs a group of parts in an assembly.

Since many industries require a wide range of sheet metal parts to house or enclose their designs, SOLIDWORKS makes a flexible design approach possible. With the help of SOLIDWORKS, one can generate base, edge, bends and many more features. How a part can be created in sheet metal and conversion will be discussed briefly in this chapter with the help of step by step tutorial and figures.

8.1 Base Flange

It creates a sheet metal part or adds material to an existing sheet metal part. To start with sheet metal

1. Click on Front Plane. Right click and select Normal To.
2. Go to Sketch toolbar and select Line and generate two lines perpendicular to each other. Right click "Select" to end the command.
3. Click on Sheet metal features toolbar and select Edge Flange/Base Flange.
4. Base Flange is created. Go to features design manager tress and edit feature and modify direction 1–50 mm.

© Springer Nature Switzerland AG 2020
K. Kumar et al., *Mastering SolidWorks*, Management and Industrial Engineering,
https://doi.org/10.1007/978-3-030-38901-7_8

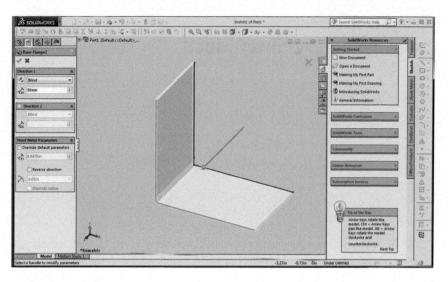

Fig. 8.1 (with permission from Dassault Systems)

5. Hit ok .

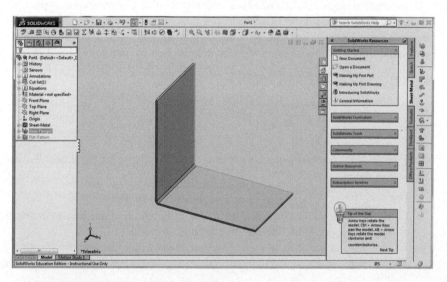

Fig. 8.2 (with permission from Dassault Systems)

6. In the Sheet metal Gauge browse and select SAMPLE TABLE-STEEL.

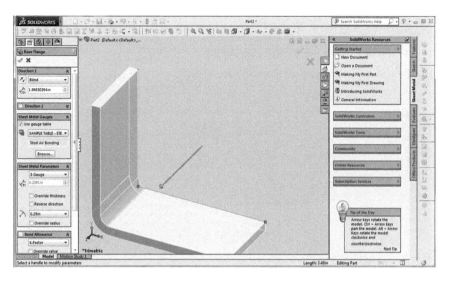

Fig. 8.3 (with permission from Dassault Systems)

7. Edit the sheet metal parameters from 3 Gauge to 18 Gauge

8. Click 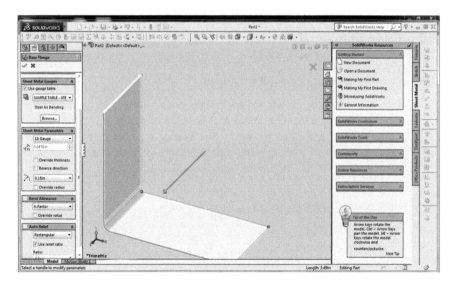 .

Fig. 8.4 (with permission from Dassault Systems)

9. After creating **base Flange/Tab** go to sheet metal and select **flatten**
   which shows the flat pattern for existing sheet metal.

10. To roll back again click the same icon

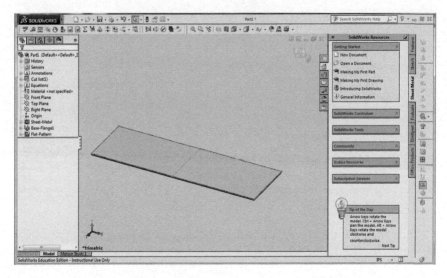

Fig. 8.5 (with permission from Dassault Systems)

8.2   Edge Flange

Adds a wall to an edge of a sheet metal part.

11. Roll back to the same sketch and go to sheet metal toolbar. Select Edge Flange.
12. In the flange parameters columns select an edge to flange.
13. Click on the location in empty space or a vertex to set the flange height.
14. Edit the parameters such as gap distance, depth direction as per the design requirement.

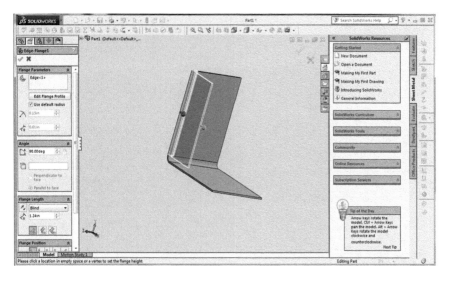

Fig. 8.6 (with permission from Dassault Systems)

15. Click ok ✓ .

16. Edge Flange created successfully.

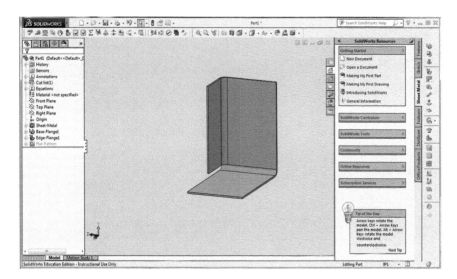

Fig. 8.7 (with permission from Dassault Systems)

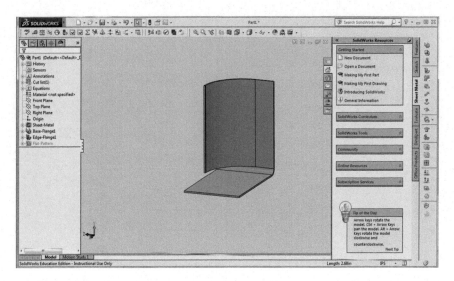

Fig. 8.8 (with permission from Dassault Systems)

17. Click on Flatten to view the flat pattern of an edge flange.

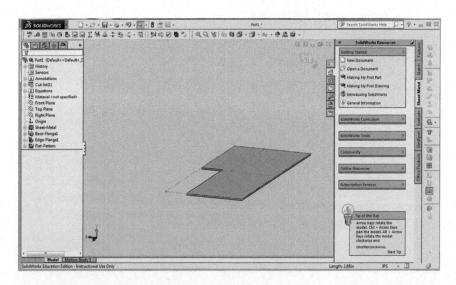

Fig. 8.9 (with permission from Dassault Systems)

8.3 Corner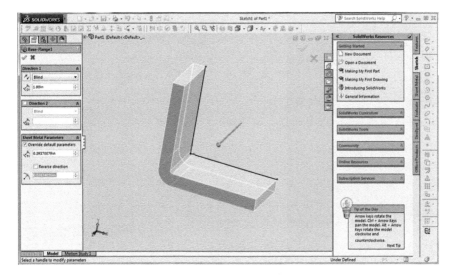

This command creates various corner treatments on a sheet metal par. To start with corners

1. Draw a sketch and select Base Flange.
2. Edit the base flange parameters and check the override default parameters to edit the thickness.

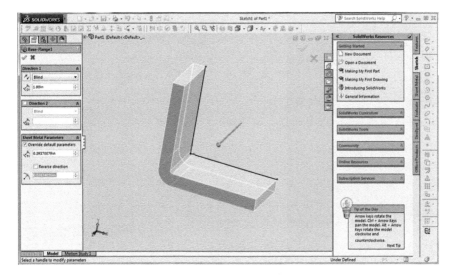

Fig. 8.10 (with permission from Dassault Systems)

3. Click ok and again go to sheet metal features toolbar and click on Edge Flange.
4. A tab opens now to create edge flange select the edges. Edge1 and Edge2 is selected.

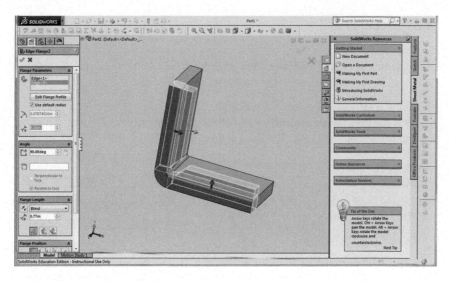

Fig. 8.11 (with permission from Dassault Systems)

5. Edit the gap distance and depth of the edge flange to be created and modify the flange height.
6. Click ok and end the command.

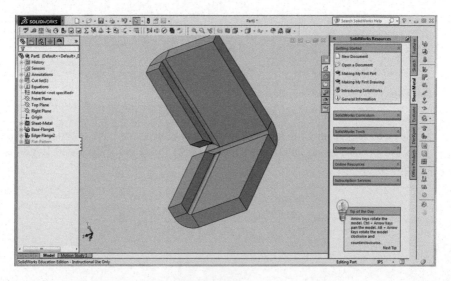

Fig. 8.12 (with permission from Dassault Systems)

7. In the sheet metal toolbar closed curve is present click on to it. Select the faces to extend (Face<1>) and in the faces to match select Face<2>.
8. Select planar corner face to extend to create a closed corner.

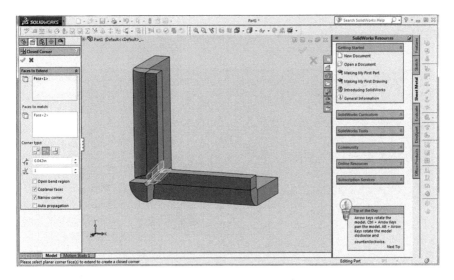

Fig. 8.13 (with permission from Dassault Systems)

9. Click ok. Closed corner created successfully.

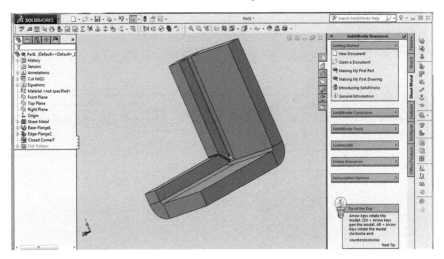

Fig. 8.14 (with permission from Dassault Systems)

8.4 Unfold

It unfolds bend in sheet metal part.

10. Click on Unfold.

11. Select the fixed face (Face<1>).
12. In Bends to Unfold section select EdgeBend3.

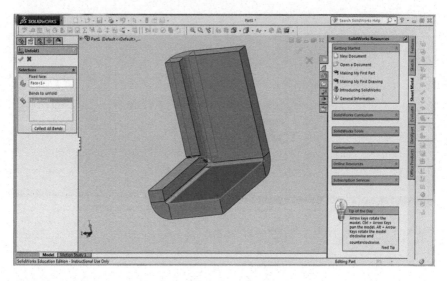

Fig. 8.15 (with permission from Dassault Systems)

13. Click ok ✓.

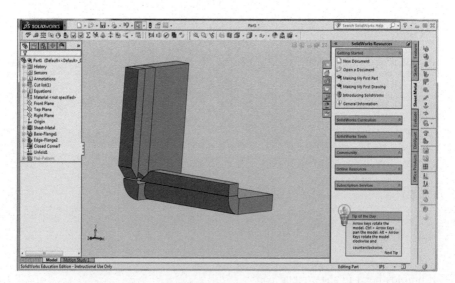

Fig. 8.16 (with permission from Dassault Systems)

8.5 Fold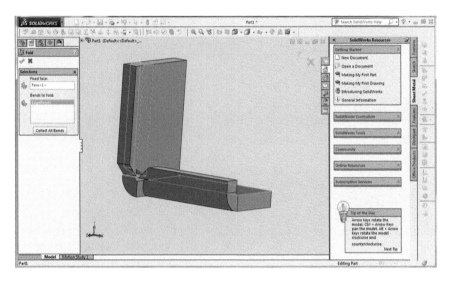

Fold flattens the bends in a sheet metal part

14. Continue with the same design sketched then go to sheet metal features toolbar and select Fold.
15. Select the fixed face as Face<1> and bends to fold as Edge Bend3.

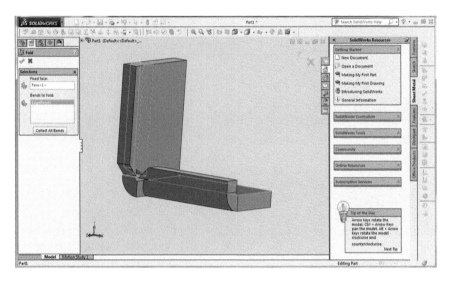

Fig. 8.17 (with permission from Dassault Systems)

16. Click ok.
17. Fold is created.

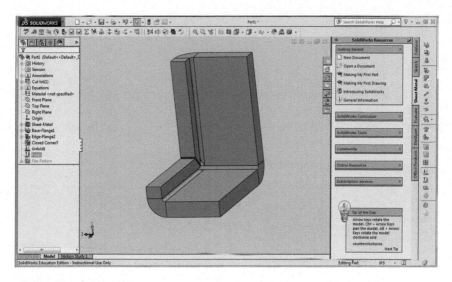

Fig. 8.18 (with permission from Dassault Systems)

8.6   Extruded Cut

It cuts a solid model by extruding a sketched profile in one or two directions. To start with this type of cut the user need to create a sketch on a part design and then perform the extruded cut

1. Taking the same sketch extruded cut is to be continued.
2. Go to sketch toolbar and select Circle.
3. Create a circle on any one of the face you require to do the cut.

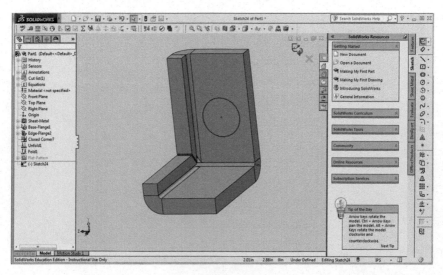

Fig. 8.19 (with permission from Dassault Systems)

4. Go to sheet metal toolbar and select cut-Extrude.
5. In the direction 1 change the depth direction from blind to Up To Next.

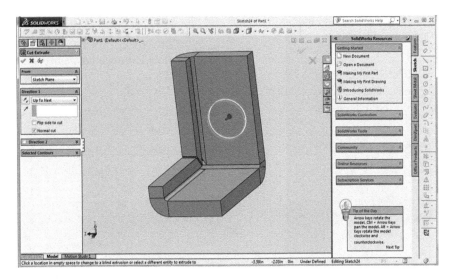

Fig. 8.20 (with permission from Dassault Systems)

6. Click ok.
7. Extruded Cut is done completely.

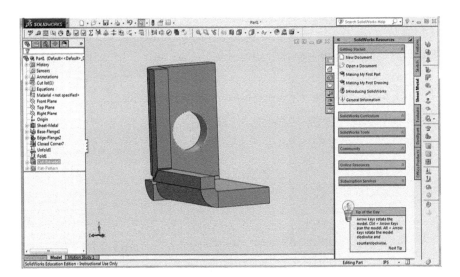

Fig. 8.21 (with permission from Dassault Systems)

8.7   Simple Hole

It creates a cylindrical hole on a planar face. To create a simple hole below given steps need to be followed

1. In the sheet metal toolbar there is a command below Extruded cut named simple Hole click on that.
2. Select Face and provide depth direction as Up To Next.
3. Edit the hole diameter.

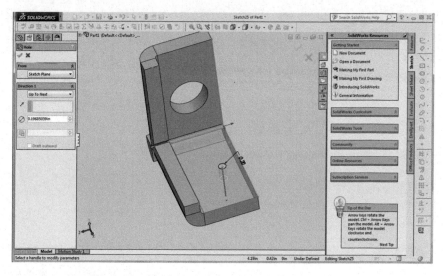

Fig. 8.22 (with permission from Dassault Systems)

4. Click ok to complete the process and end the command.

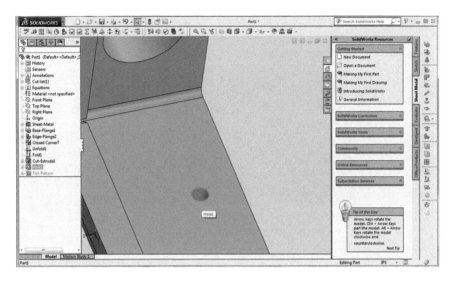

Fig. 8.23 (with permission from Dassault Systems)

8.8 Miter Flange

It adds a series of flanges to one or more edges of a sheet metal part. To create miter flange

1. Select a plane and create a center rectangle.
2. Go to sheet metal select Base Flange. Edit the thickness.
3. Click ok.

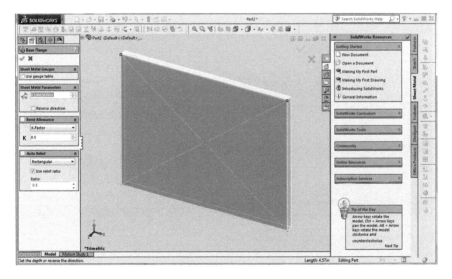

Fig. 8.24 (with permission from Dassault Systems)

4. Select the face and right click to make it Normal To the screen.

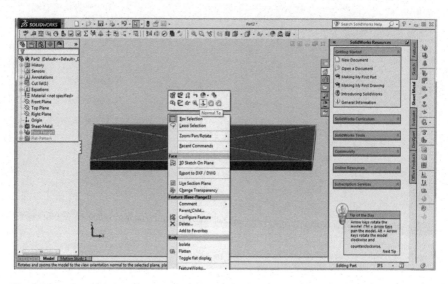

Fig. 8.25 (with permission from Dassault Systems)

5. Go to sketch toolbar section and draw two lines perpendicular to each other.
6. Now click on Top plane.

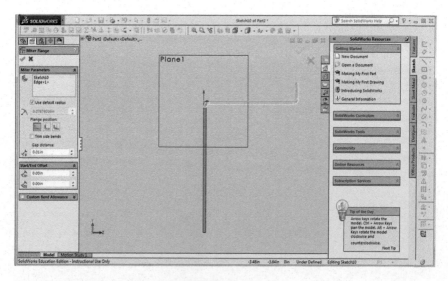

Fig. 8.26 (with permission from Dassault Systems)

7. Select Miter Flange and now click to select the miter parameters (Sketch and Edge<1>). Three flange positions are available in miter flange which are material inside, material outside and bend outside. First one is considered here.

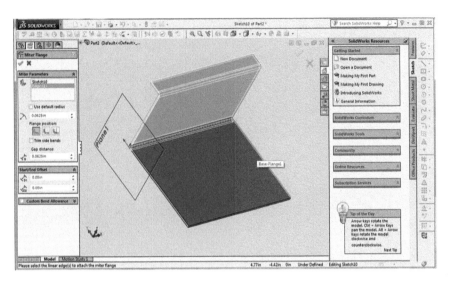

Fig. 8.27 (with permission from Dassault Systems)

8. Select the linear edges to attach the miter flange. Add all the left out edges to generate miter flange.

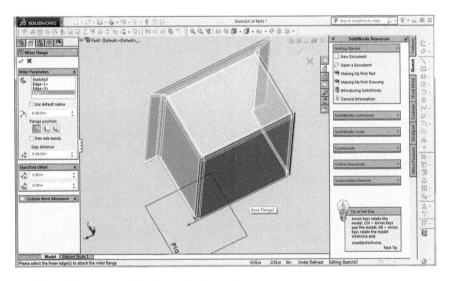

Fig. 8.28 (with permission from Dassault Systems)

9. Click Ok to complete the feature.

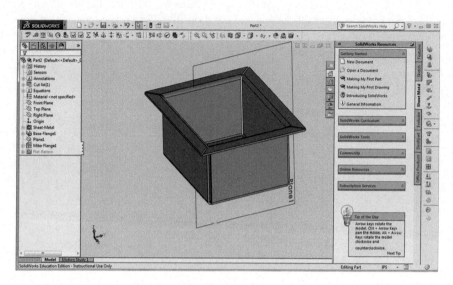

Fig. 8.29 (with permission from Dassault Systems)

Chapter 9
Advanced Features in Sheet Metal

In this chapter, some of the advanced features of sheet metal operations are used which helps in addition of two bends, curling of the edges, attachment of two edges with help of gusset and also of creating gap between the edges. All the above said can be done in sheet metal which is discussed one by one in this chapter. Some of the features are jog, hem, sketched bend, Rip, sheet metal gusset and sketched bend.

9.1 Jog

It adds two bends from a sketched line in a sheet metal part. It can be done in following way.

It adds two bends from a sketched line in a sheet metal part. It can be done in following way.

1. Select a plane (Front Plane).
2. Create a center rectangle and click ok.
3. Go to Sheet metal toolbar and click base flange and edit the thickness of the flange.
4. Click ok. Base flange created.
5. Click on the plane and then select Normal To the screen.

© Springer Nature Switzerland AG 2020
K. Kumar et al., *Mastering SolidWorks*, Management and Industrial Engineering,
https://doi.org/10.1007/978-3-030-38901-7_9

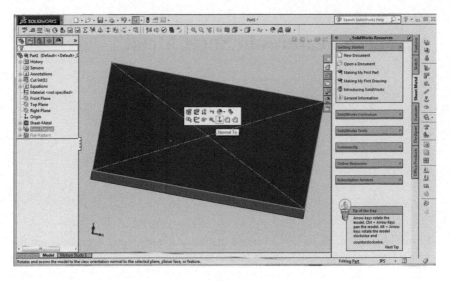

Fig. 9.1 (with permission from Dassault Systems)

6. Go to sketch and generate line and right-click to select and then end the command.

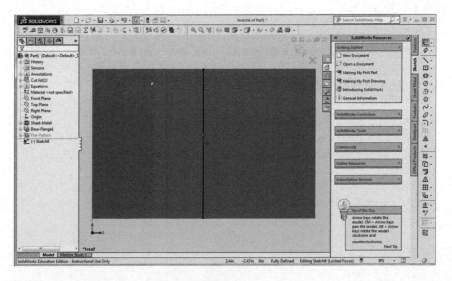

Fig. 9.2 (with permission from Dassault Systems)

7. Go to sheet metal and select Jog. Select a fixed face as <face 1> and specify the planar face to be fixed when creating the bend. Uncheck the default radius and give edit as per the dimension required. Edit the offset distance.

8. Three dimension positions are given in jog which is outside offset, inside offset, overall dimension. Click to see the difference in the position. Four jog positions are given: bend centerline, material inside, material outside and bend outside. Bend inside is selected. Edit the jog angle.

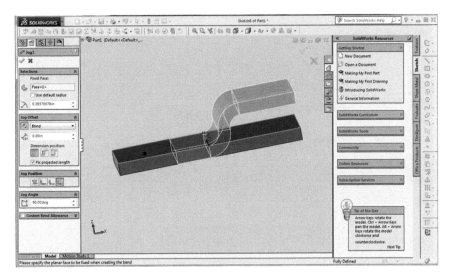

Fig. 9.3 (with permission from Dassault Systems)

9. Click ok.

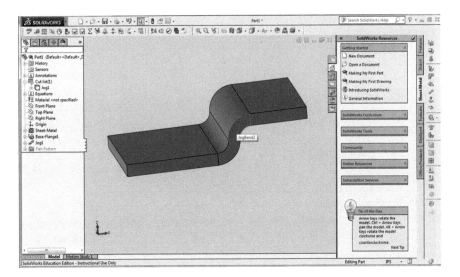

Fig. 9.4 (with permission from Dassault Systems)

10. Click on Jog and select edit feature.
11. Edit angle to 120° and reverse the direction.
12. Edit the jog position to jog inside and offset it to zero.

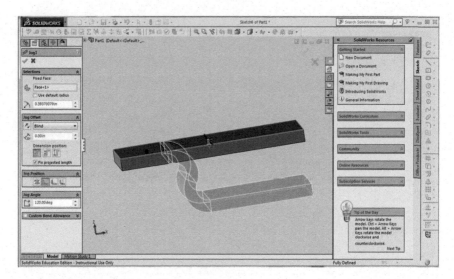

Fig. 9.5 (with permission from Dassault Systems)

13. Click ok.

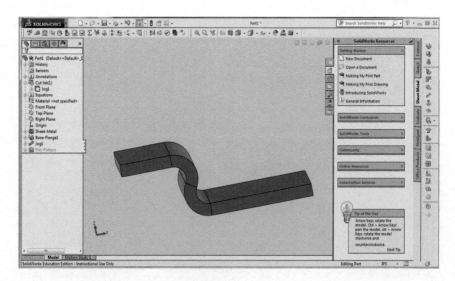

Fig. 9.6 (with permission from Dassault Systems)

9.2 Hem

Hem is used to curl edges of a sheet metal part. To generate a hem follow the given steps provided below.

1. Select a plane (Front Plane).
2. Sketch a center rectangle and extrude it with the help of extruded boss/base.
3. Convert it to sheet metal by selecting it from sheet metal feature toolbar.
4. Click Hem.
5. Select edge 1 and edge 2 and edit hem depth as material inside and bend outside.
6. Different types and size are given to select it as per the requirement and see the difference.

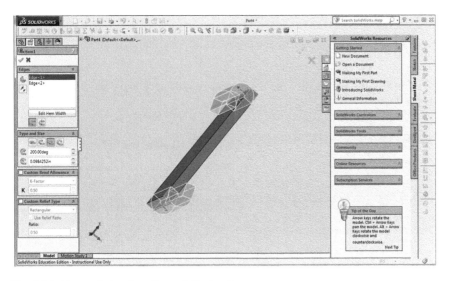

Fig. 9.7 (with permission from Dassault Systems)

7. Click ok.

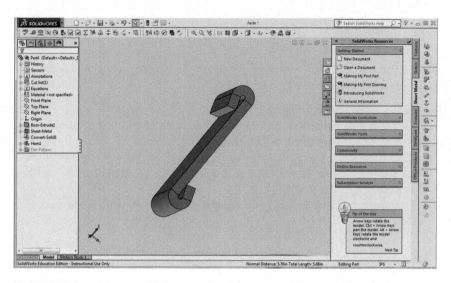

Fig. 9.8 (with permission from Dassault Systems)

8. Edit feature and modify the length and gap distance and click tear drop. Click ok.

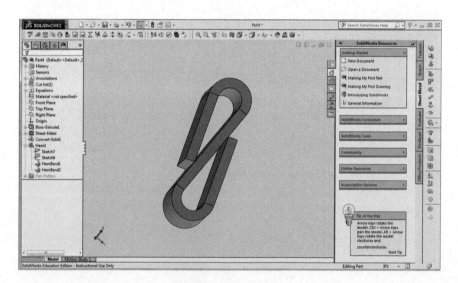

Fig. 9.9 (with permission from Dassault Systems)

9. Select only one edge and type is open.

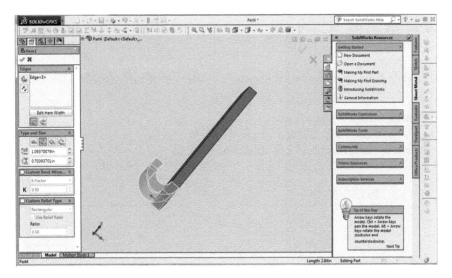

Fig. 9.10 (with permission from Dassault Systems)

10. Click ok.

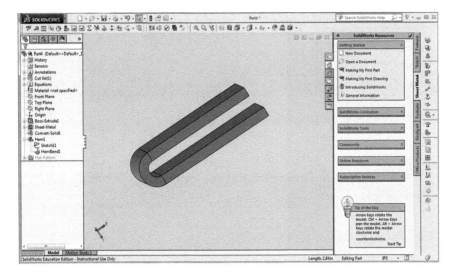

Fig. 9.11 (with permission from Dassault Systems)

9.3 Sketched Bend

It adds a bend from a selected sketch in a sheet metal part. Sketched bend can be created in below given steps

1. Select a plane (Front Plane).
2. Go to sketch and create a corner rectangle.
3. Select base flange and edit its depth. Click ok.
4. Go to line and create an inclined line on the plane selected.

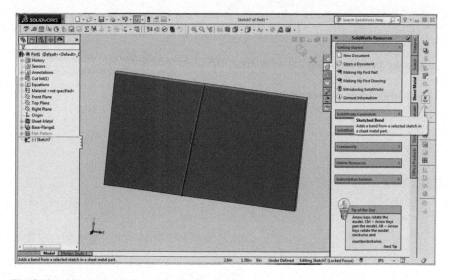

Fig. 9.12 (with permission from Dassault Systems)

5. In the sheet metal toolbar, click Sketched bend. Four bend positions are given : bend centerline, material inside, material outside and bend outside. Select bend centerline.
6. Click ok.

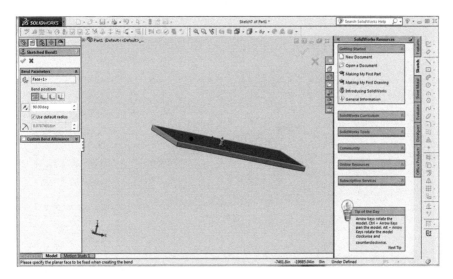

Fig. 9.13 (with permission from Dassault Systems)

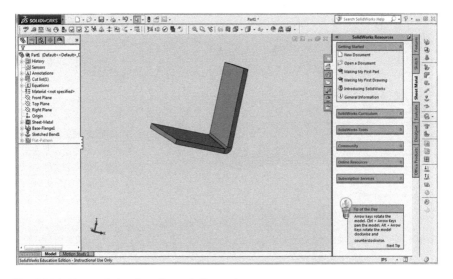

Fig. 9.14 (with permission from Dassault Systems)

7. Sketched bend created.
8. In the feature manager design tree, click on sketched bend and edit feature.
9. Edit bend position as material inside and uncheck the use default radius.

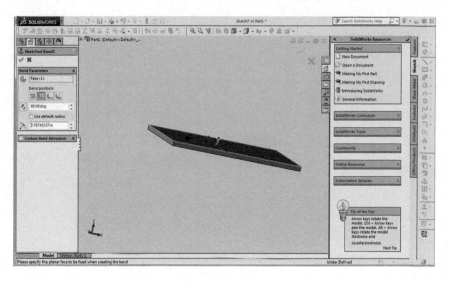

Fig. 9.15 (with permission from Dassault Systems)

10. Click ok to complete the feature successfully.

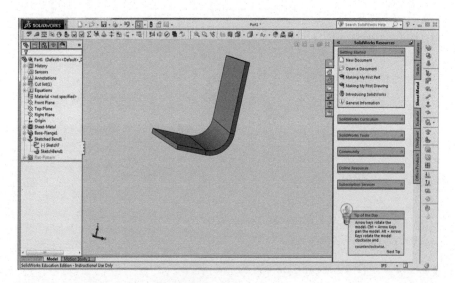

Fig. 9.16 (with permission from Dassault Systems)

9.4   Forming Tool

It creates a inverse dent on a sheet metal part. Forming tool can be created in following steps:

1. Select a plane and create a rectangle. With the help of cursor select two perpendicular lines and make it construction geometry. Click ok.

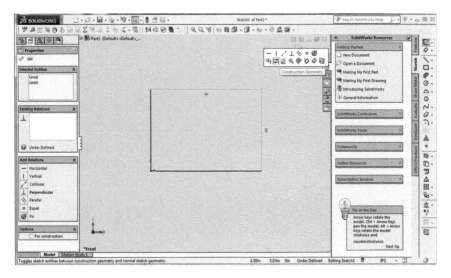

Fig. 9.17 (with permission from Dassault Systems)

2. Click on arc to create a 3 point arc and make both the selected lines equal.

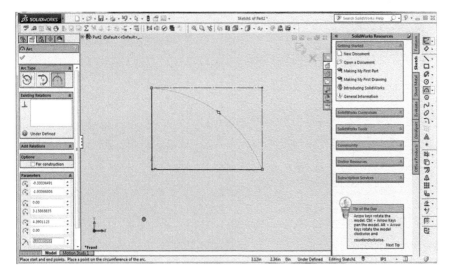

Fig. 9.18 (with permission from Dassault Systems)

3. Select the face and go to sketch to convert entities.

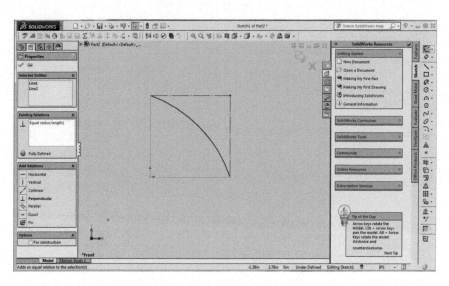

Fig. 9.19 (with permission from Dassault Systems)

4. Extrude it.

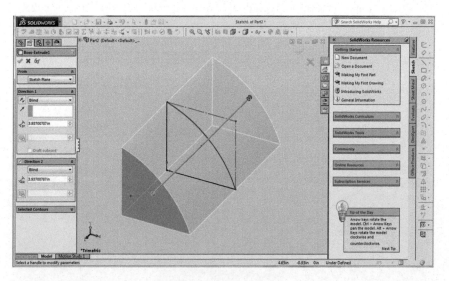

Fig. 9.20 (with permission from Dassault Systems)

5. Go to features toolbar now and click on Revolve. Select line 8 and edit the direction angle.

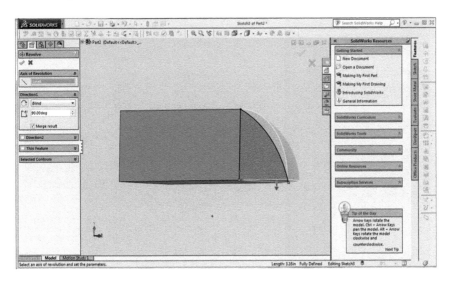

Fig. 9.21 (with permission from Dassault Systems)

6. Click on right plane and in sketch toolbar select the mirror option is there select it.

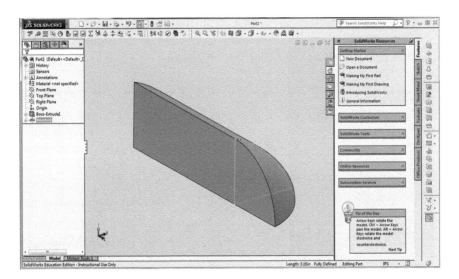

Fig. 9.22 (with permission from Dassault Systems)

7. Select mirror face/plane and click on features to mirror.

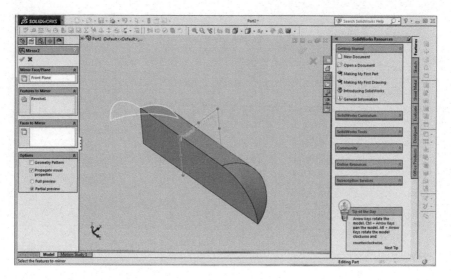

Fig. 9.23 (with permission from Dassault Systems)

8. Click ok to complete mirror. Select right plane and sketch center rectangle as shown in below figure. Reverse the direction and make it blind. Click on fillet by selecting the edges.

9. Click on top plane sketch by making it normal to the screen. Create corner rectangle. Click ok. Go to sheet metal toolbar and select extruded cut. Select in both the direction and make the cut through all.

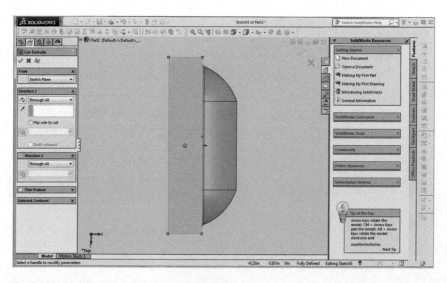

Fig. 9.24 (with permission from Dassault Systems)

10. Click ok.

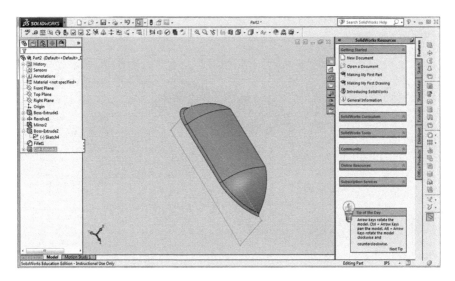

Fig. 9.25 (with permission from Dassault Systems)

11. Now click on forming tool. Select stopping face as Face<1> and faces to be removed as Face<2>.

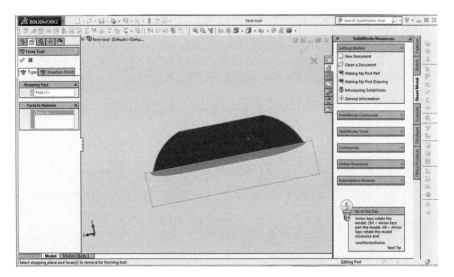

Fig. 9.26 (with permission from Dassault Systems)

12. Click ok.
13. Forming tool created successfully.

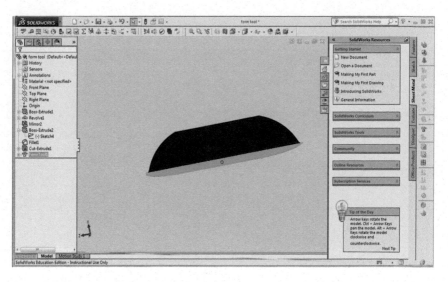

Fig. 9.27 (with permission from Dassault Systems)

9.5   Cross Break

Adds a cross break feature to a selected face.

1. Select a plane (front plane).
2. Click on base flange from sheet metal toolbar then define a sketch by creating a center rectangle. Exit the sketch. Base flange will be created.
3. Now click on Edge flange and select the edges, i.e., Edge 1 and Edge 2. Edit the flange height and the gap distance as per the dimension.

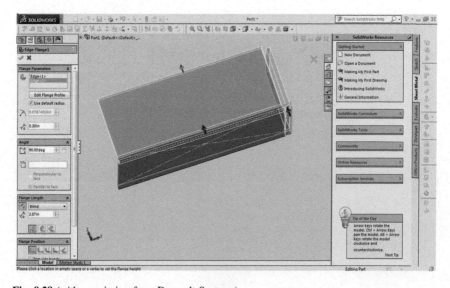

Fig. 9.28 (with permission from Dassault Systems)

4. Click ok. Edge flange will be created at both the edges. In the sheet metal toolbar, click on cross break.
5. Select the cross break parameters which should be a face.

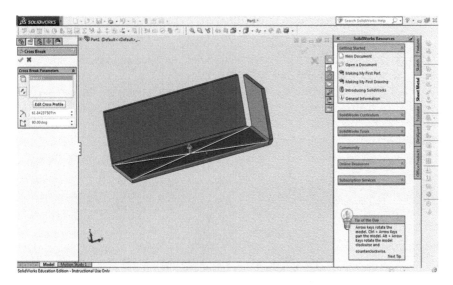

Fig. 9.29 (with permission from Dassault Systems)

6. Click ok. Cross break will be generated successfully.

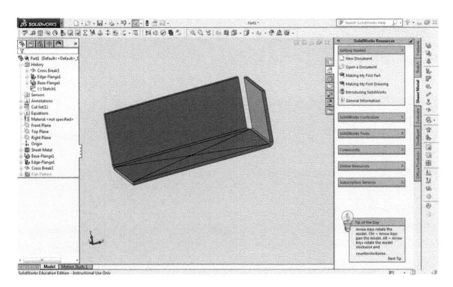

Fig. 9.30 (with permission from Dassault Systems)

9.6 Rip

It creates a gap between two edges in a sheet metal part. To create a rip in sheet metal below given steps need to be followed.

1. Select a plane. Create a rectangle and extrude it.
2. Go to features toolbar and select shell. Select the face and create a shell.
3. In the sheet metal part, click rip.
4. Select the rip parameters by selecting all the edges by moving the part sketched with the help of cursor.

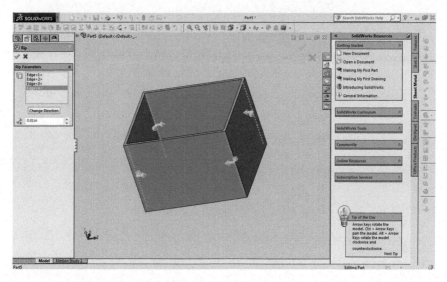

Fig. 9.31 (with permission from Dassault Systems)

5. Click ok. Rip will be generated successfully.

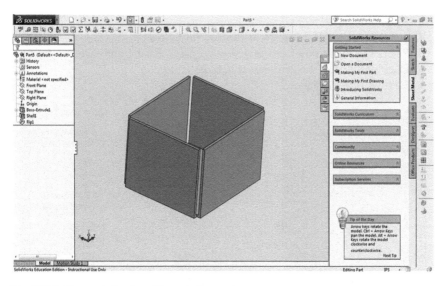

Fig. 9.32 (with permission from Dassault Systems)

9.7   Sheet Metal Gusset

It adds a gusset/rib across a bend of a sheet metal body. To create a sheet metal gusset in SOLIDWORKS below given steps need to be followed.

1. Select a plane and create two lines perpendicular to each other.
2. Click on base flange and edit the thickness to increase it.

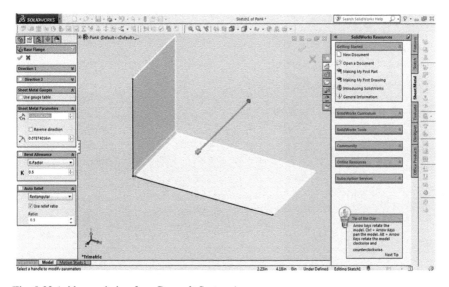

Fig. 9.33 (with permission from Dassault Systems)

3. Click ok.
4. Select sheet metal gusset from sheet metal toolbar.
5. Select the supporting faces to generate gusset. Face 1 and face 2 is selected.
6. Click on the edges where gusset is to be created.
7. Edit the offset distance.

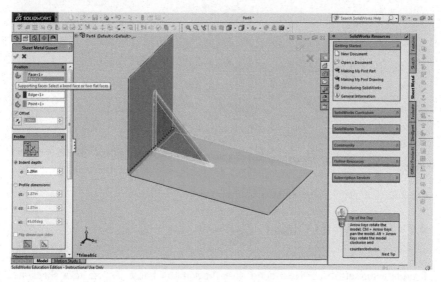

Fig. 9.34 (with permission from Dassault Systems)

8. Increase the indent depth d.

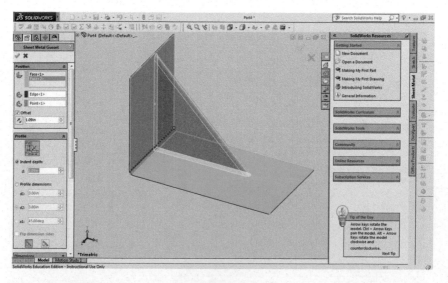

Fig. 9.35 (with permission from Dassault Systems)

9. Move the cursor down to check other options. Dimensions can be seen where indent width can be provided and thickness can be increased.

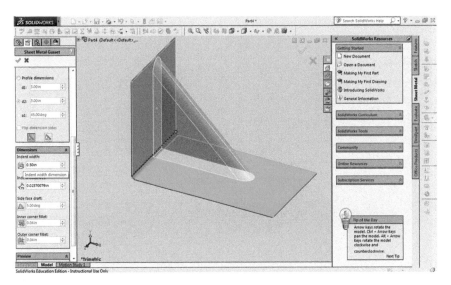

Fig. 9.36 (with permission from Dassault Systems)

10. Click ok.
11. Sheet metal gusset created.

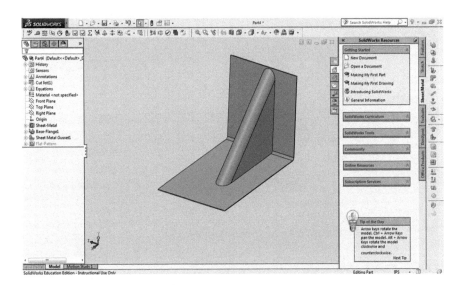

Fig. 9.37 (with permission from Dassault Systems)

9.8   No Bends

It rolls back all bends in the sheet metal part. A change in the sketch can be seen
after clicking or selecting the icon.

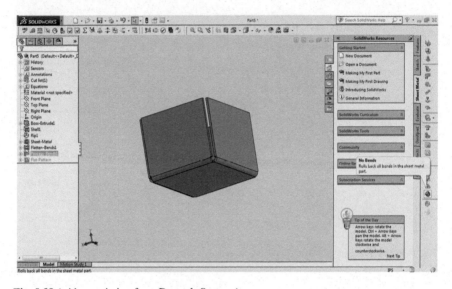

Fig. 9.38 (with permission from Dassault Systems)

Edges which are bend are converted into flat sheet in the figure given below.

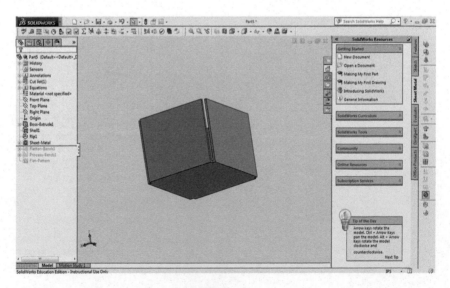

Fig. 9.39 (with permission from Dassault Systems)

Chapter 10
Assembly

Assembly is a universal joint made by combining individual components or parts. There are number of stages in the assembly process.

1. **Creation of a new assembly**: New assemblies are created using the same method as new parts.
2. **Adding the first component**: Components can be added in several ways. They can be dragged and dropped from an open part window or opened from a standard browser.
3. **Position of the first component**: The initial component added to the assembly is automatically fixed as it is added. Other components can be positioned after they are added.
4. **Feature Manager Design tree and symbols**: The feature manager includes many symbols, prefixes, suffix that provides information about the assembly and the components in it.
5. **Mating components to each other**: Mates are used to position and orient components with reference to each other. Mates remove degree of freedom from the components.
6. **Sub-assemblies**: Assemblies can be created and inserted into the current assembly. They are considered sub-assembly components.

New assemblies can be created directly or be made from an open part or assembly. The new assembly contains an origin, the three standard reference planes and a special feature.

Use the make assembly from part/assembly option to generate a new assembly from an open part. The part is used as a reference component in the new assembly and is fixed in space.

© Springer Nature Switzerland AG 2020   227
K. Kumar et al., *Mastering SolidWorks*, Management and Industrial Engineering,
https://doi.org/10.1007/978-3-030-38901-7_10

Where to find it?

1. Click **make Assembly from Part/Assembly** option on the standard toolbar.
2. Click **File, Make Assembly from Part**.

To start with assembly first of all we need to create parts so that they can be assembled with the help of assembly.

1. Go to **part** design.
2. Select plane (**front Plane**) which should be **normal to** the screen.
3. Go to sketch toolbar and click on **circle**.
4. Create a circle of 8 mm and extrude it.
5. Select one face of the circle to sketch another circle of 3 mm and extrude it.

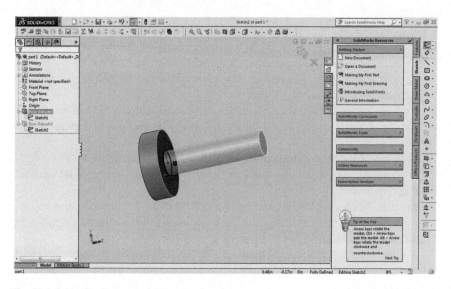

Fig. 10.1 (with permission from Dassault Systems)

6. Click **ok** 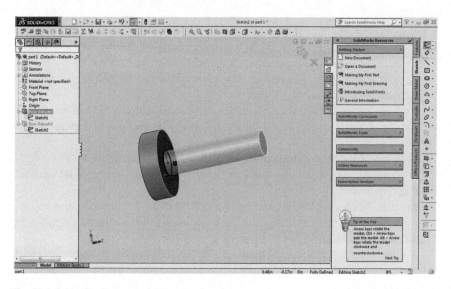.
7. Go to **File** and **Save as**.

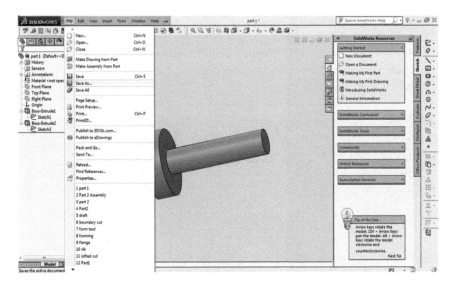

Fig. 10.2 (with permission from Dassault Systems)

8. **Save** the component as **Part 1**.
9. Click on **Save**.

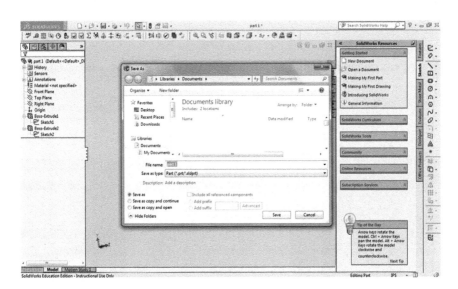

Fig. 10.3 (with permission from Dassault Systems)

10. Again click on **New Document**.
11. Select **Part** and click **ok**.
12. Select a plane and create a sketch with the help of **circle** command.
13. Create two circles one of 8 mm and other of 3 mm.
14. **Extrude it**.

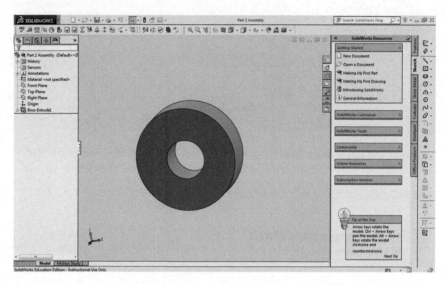

Fig. 10.4 (with permission from Dassault Systems)

15. **Save** the component named as **Part 2**.
16. After creating both the parts click on **New Document** and select **Assembly**.

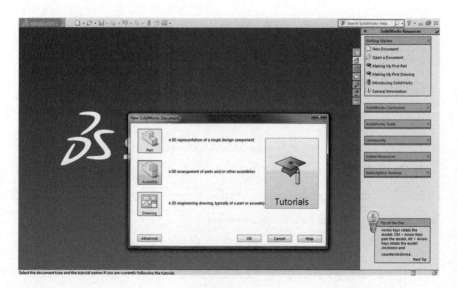

Fig. 10.5 (with permission from Dassault Systems)

17. Or else go to **file** and click **make assembly from part**.

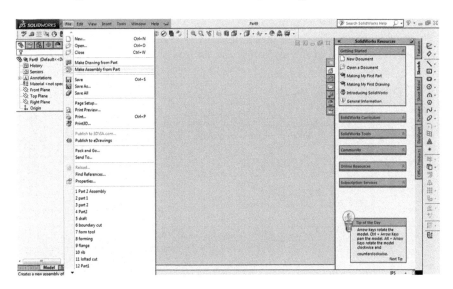

Fig. 10.6 (with permission from Dassault Systems)

18. Assembly tab will be opened to begin with.

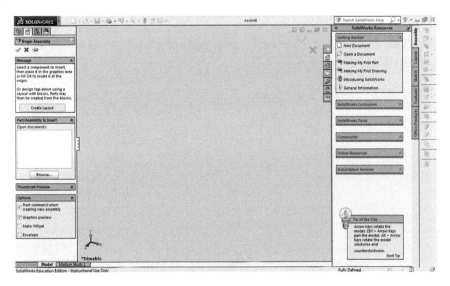

Fig. 10.7 (with permission from Dassault Systems)

19. In the left-hand side, **open documents** can be seen so click on **Browse** and open the first part to insert into the assembly. The initial component added to

the assembly is, by default, fixed. Fixed components cannot be moved and are locked into place wherever they fall when you insert them into the assembly. By using the cursor during placement, the components origin is at assembly origin position. This also means that the reference planes of the component match the planes of the assembly, and the component is fully defined.

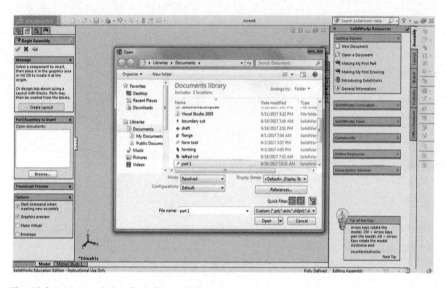

Fig. 10.8 (with permission from Dassault Systems)

20. Open the **Part 1** and browse for other part and click on the graphics window screen to settle it.

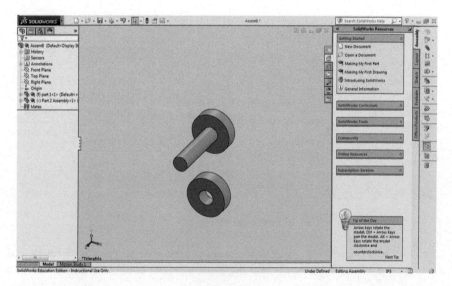

Fig. 10.9 (with permission from Dassault Systems)

21. In the **assembly toolbar**, there is an icon from where **insertion of the component/new part** can be done as shown in the figure given below.

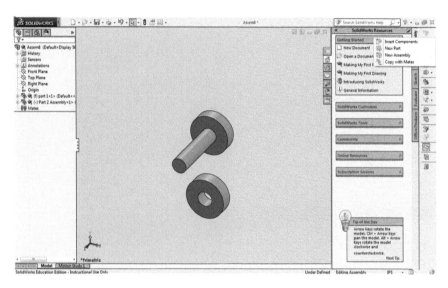

Fig. 10.10 (with permission from Dassault Systems)

22. There are six degrees of freedom for any component that is added to the assembly before it is mated or fixed: translation along the *X*, *Y*, and *Z* axes and rotation around those same axes. How a component is able to move in the assembly is determined by its degree of freedom. The Fix and the Mate options are used to remove degrees of freedom.

23. There are different types of **mates** present that is standard mates, mechanical mates, and advanced mates. The mating relationships in assemblies are grouped together into a Mate folder named Mates. A mate group is a collection of mates that get solved in order in which they are listed. All assemblies will have mate group.

24. To **mate the components** select the faces of the part and click **coincident**.

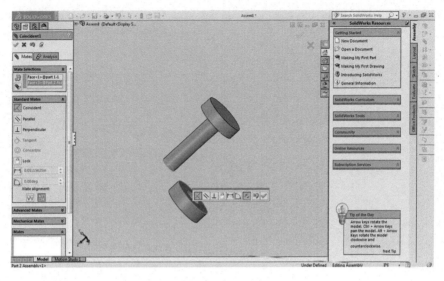

Fig. 10.11 (with permission from Dassault Systems)

25. It will be seen in the figure that part becomes **coincident** ⟨ to each other.

26. After that select the faces and click on **parallel** ⟍ to make the components parallel to each other.

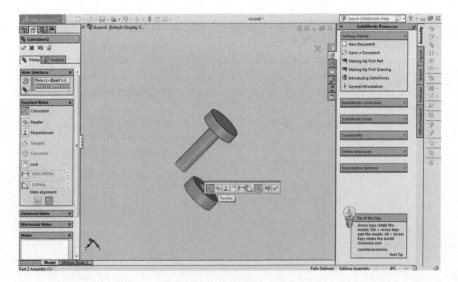

Fig. 10.12 (with permission from Dassault Systems)

27. Clicking on the icon will make the parts parallel to each other.

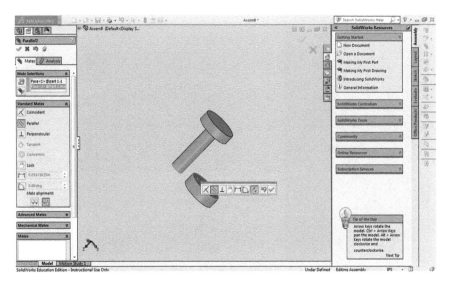

Fig. 10.13 (with permission from Dassault Systems)

28. Next is **perpendicular** ⊥. Selecting the perpendicular icon will make the components perpendicular to each other.

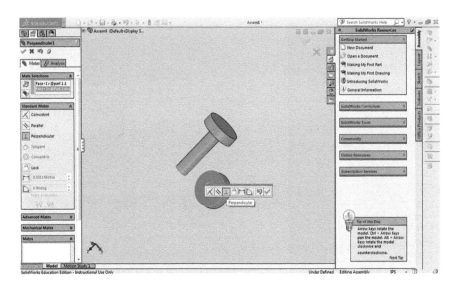

Fig. 10.14 (with permission from Dassault Systems)

29. Components can be locked by clicking on **lock** icon ![lock icon]. Undo can be done by clicking on ![undo icon] icon.

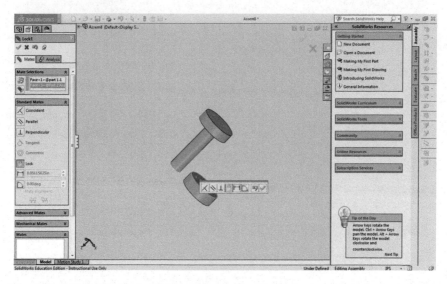

Fig. 10.15 (with permission from Dassault Systems)

30. Now to insert Part 2 in Part 1 the user has to make it coincident to each other. So select **Edge<1>** of Part 1 and **Face<2>** of Part 2 and it will become **co-incident** ![icon].

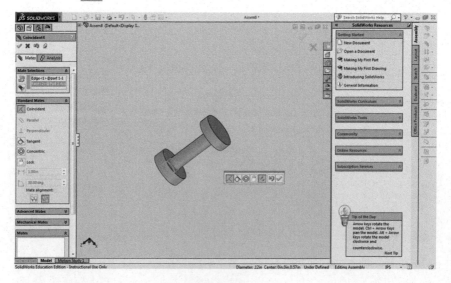

Fig. 10.16 (with permission from Dassault Systems)

31. **Edge<1>** of Part 1 and **Edge<2>** of Part 2 will be made **concentric** to each other.

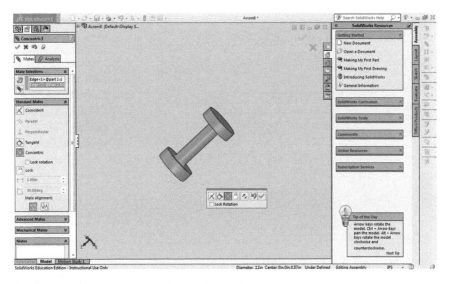

Fig. 10.17 (with permission from Dassault Systems)

32. Select a **Move Component** 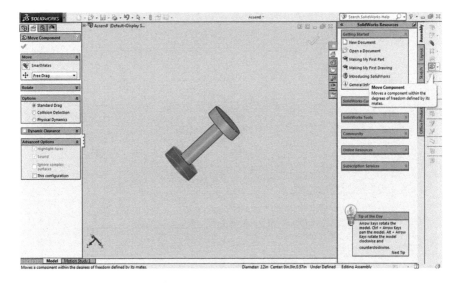 which moves the component within the degrees of freedom defined by its mates.

Fig. 10.18 (with permission from Dassault Systems)

33. Now drag Part 2 to insert it into Part 1.

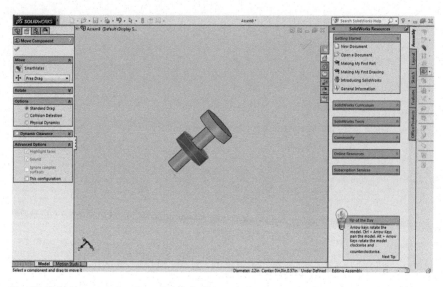

Fig. 10.19 (with permission from Dassault Systems)

34. From the assembly toolbar, select **bill of materials** and click below the assembly. Bill of materials tells about the part number its description and quantity.

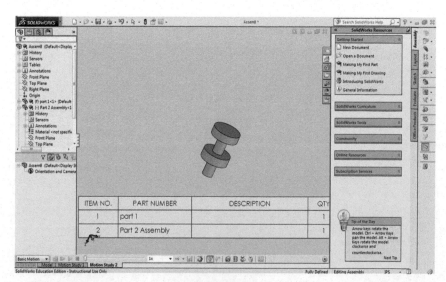

Fig. 10.20 (with permission from Dassault Systems)

Chapter 11
Assembly Continued

In the last chapter, assembly was done for a circular component utilizing the centerline which takes care of dimensions with reference to both x-axis and y-axis. In this chapter, a non-circular component would be assembled wherein distances from both the axes have to be defined. The procedure is as described below.

1. Click on **File** and select **New...**.

Fig. 11.1 (with permission from Dassault Systems)

2. A **New SOLIDWORKS Document** window will pop up. Click on **Part** and Click **OK**.

© Springer Nature Switzerland AG 2020
K. Kumar et al., *Mastering SolidWorks*, Management and Industrial Engineering,
https://doi.org/10.1007/978-3-030-38901-7_11

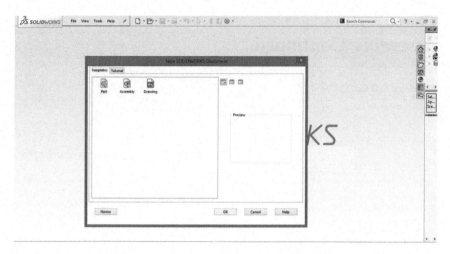

Fig. 11.2 (with permission from Dassault Systems)

3. Right-click on **Front Plane** and **Sketch**. **Front Plane** will be displayed on the screen.

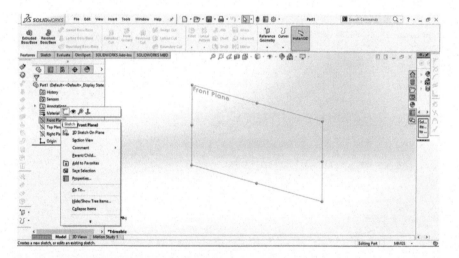

Fig. 11.3 (with permission from Dassault Systems)

4. Click on **Sketch** and select the **Rectangle** option.

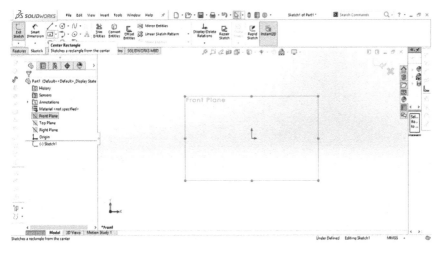

Fig. 11.4 (with permission from Dassault Systems)

5. Select the **Center Rectangle** option under the **Rectangle window**. **Left-click** on the sketcher screen and then **drag the mouse** and then **terminate the Left-click** to generate a rectangle.

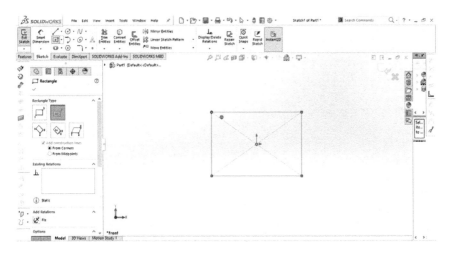

Fig. 11.5 (with permission from Dassault Systems)

6. Click on **Sketch** and select **Smart Dimensions** for specifying the dimensions of the rectangle. Exit the sketcher window by clicking 🔲 button.

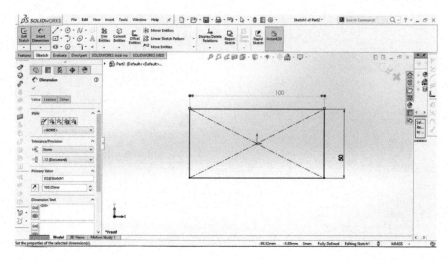

Fig. 11.6 (with permission from Dassault Systems)

7. Click on **Features** and then select the **Extruded Boss/Base option**. Enter the extrude depth of 10 mm under the **Direction 1 option**. Clicking the ✓ button creates a 3D rectangle.

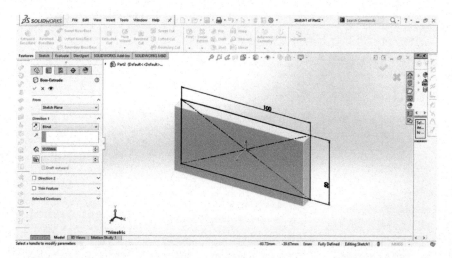

Fig. 11.7 (with permission from Dassault Systems)

8. Orient the generated cuboid on its front plane. Click on **Sketch** and select **Rectangle**. Under the rectangle window, select **Center Rectangle** to create a rectangle.
9. Click on **Smart Dimensions** for the dimensioning of the created rectangle. Exit the sketcher window by clicking ⬚ button.

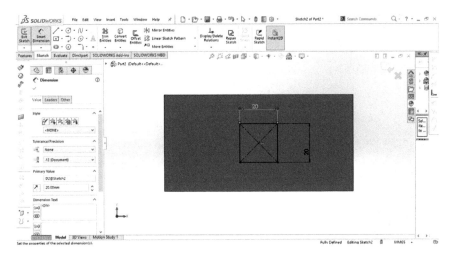

Fig. 11.8 (with permission from Dassault Systems)

10. Click on **Features** and select **Extruded Cut** option.

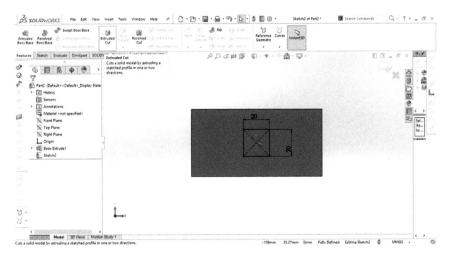

Fig. 11.9 (with permission from Dassault Systems)

11. Enter the **depth of extruded cut** in the **Direction 1** filed to create a cut of the desired depth. Click on [icon] to terminate the extruded cut option.

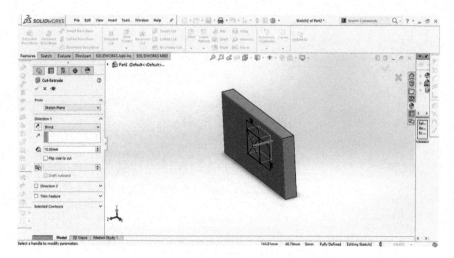

Fig. 11.10 (with permission from Dassault Systems)

12. Click on **File** and select **Save As…** to save the part created.

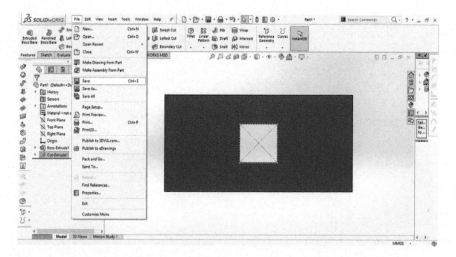

Fig. 11.11 (with permission from Dassault Systems)

13. Type the file name as **Part 1** in **File name** filed. Click on **Save** to save the created part.

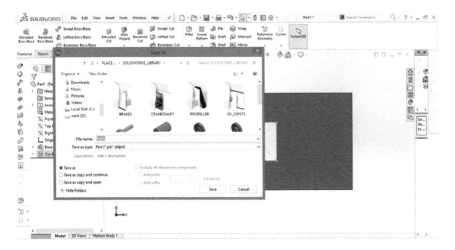

Fig. 11.12 (with permission from Dassault Systems)

14. Click on **File** and select **New…**.
15. A **New SOLIDWORKS Document** window will pop up. Click on **Part** and Click **OK**.
16. Right-click on **Front Plane** and **Sketch**. **Front Plane** will be displayed on the screen.
17. **Repeat steps 4, 5, and 6** to create another rectangle with the dimensions as shown in the figure.

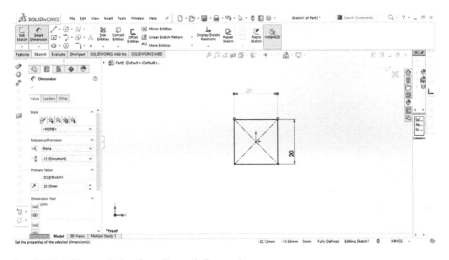

Fig. 11.13 (with permission from Dassault Systems)

18. Select **Features** and click on **Extruded Boss/Base**. Select the **Blind** option under Direction 1 and then provide the **depth of extrude** as **10 mm**. **Save** the part with file name **Part 2**.

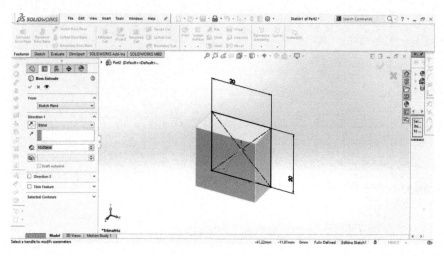

Fig. 11.14 (with permission from Dassault Systems)

19. Next the assembly of the created parts will be done. Click on **File** and then select **New**. Then select **Assembly module** in the **New SOLIDWORKS Document** window and click **OK**.

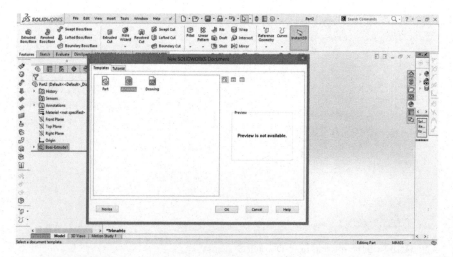

Fig. 11.15 (with permission from Dassault Systems)

20. The assembly user interface is shown in the figure. To add parts for the assembly click on **Browse** and then select the created parts from the directory or click on **Insert Components** to add the parts to the assembly interface.

Fig. 11.16 (with permission from Dassault Systems)

Fig. 11.17 (with permission from Dassault Systems)

21. The parts added to the assembly interface will appear as shown in the figures.

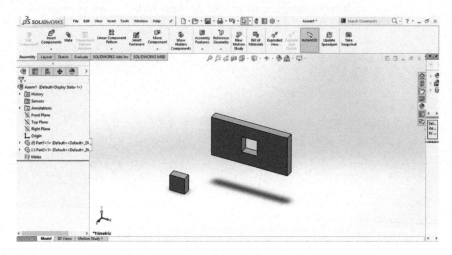

Fig. 11.18 (with permission from Dassault Systems)

22. Click on **Mate**. The **Mate** window will open up which comprises of the **Mate Selection** field and **Standard Mates**.

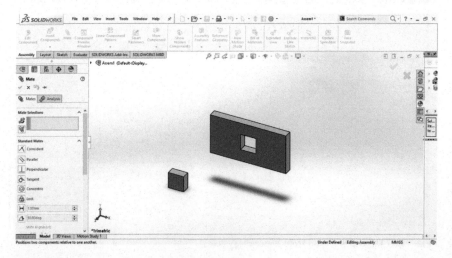

Fig. 11.19 (with permission from Dassault Systems)

23. Click on **Move Components** to orient the parts if required. Select the side faces of the parts created for the **Mate Selection** field. Next click on **Coincident** under **Standard Mates**.

Fig. 11.20 (with permission from Dassault Systems)

24. Select the two overlooking faces of the created parts as the two faces for the **Mate Selections** field. Next click on **Coincident** under the **Standard Mates option**. Click on ✓ to terminate the process.

Fig. 11.21 (with permission from Dassault Systems)

25. The assembly created is shown in the Fig. 11.22.

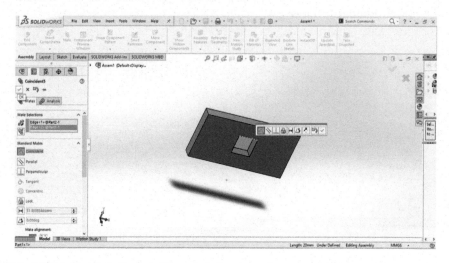

Fig. 11.22 (with permission from Dassault Systems)

26. Save the created assembly.

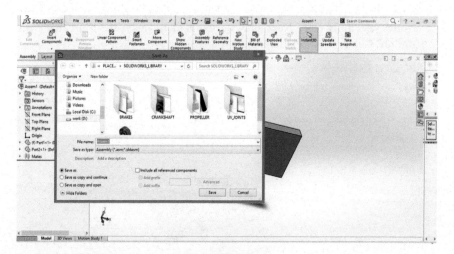

Fig. 11.23 (with permission from Dassault Systems)

Chapter 12
Basics of Drawing

Drawing consists of one or more views generated from a part or assembly. A part or assembly is saved before creating its associated drawing. To create a drawing follows the given steps:

1. Create a rib or open any part which is saved in your documents.
2. Click on the **view orientation** and select Top view.

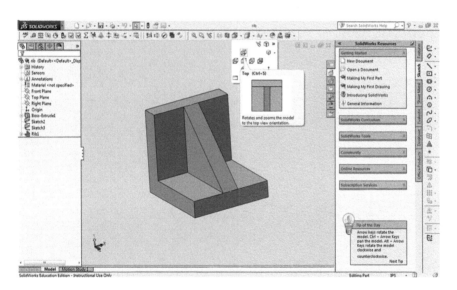

Fig. 12.1 (with permission from Dassault Systems)

© Springer Nature Switzerland AG 2020
K. Kumar et al., *Mastering SolidWorks*, Management and Industrial Engineering,
https://doi.org/10.1007/978-3-030-38901-7_12

3. Clicking on **Top view** will show the top view of the part designed.

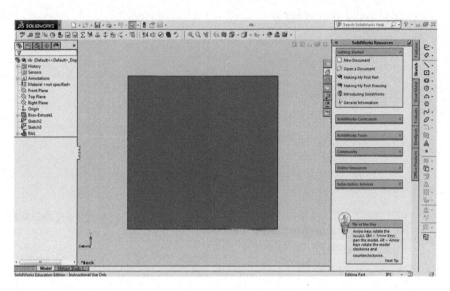

Fig. 12.2 (with permission from Dassault Systems)

4. Secondly, select the **back view**. A rectangle can be seen from the back view of the part.

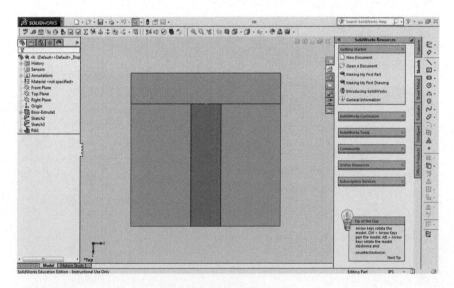

Fig. 12.3 (with permission from Dassault Systems)

5. Now select the **left orientation** to view the left side of the part.

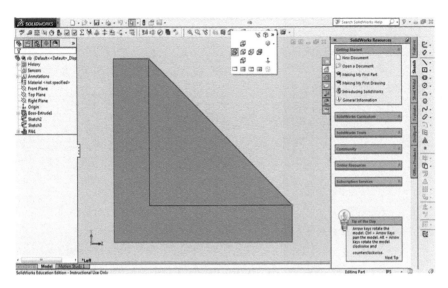

Fig. 12.4 (with permission from Dassault Systems)

6. Next is **Front view**.

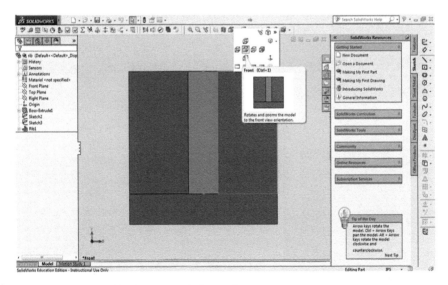

Fig. 12.5 (with permission from Dassault Systems)

7. So all the view can be seen by clicking on the **view orientation**.
8. Go to **New** menu and select **Make Drawing from Part/Assembly**. It will create a new drawing of the current part/assembly.

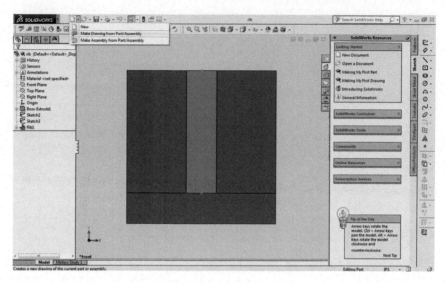

Fig. 12.6 (with permission from Dassault Systems)

9. A sheet format will open, select any one of them. By default, A (ANSI) Landscape will be selected. Click ok.

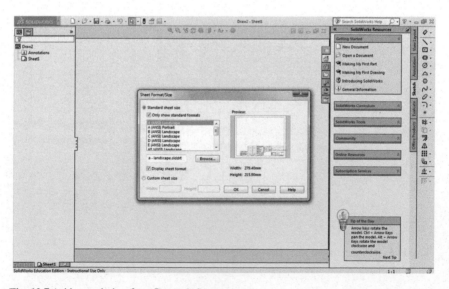

Fig. 12.7 (with permission from Dassault Systems)

10. **Sheet Format 1** will be opened and in the right-hand side **View Palette** will be available with all the views of the part.

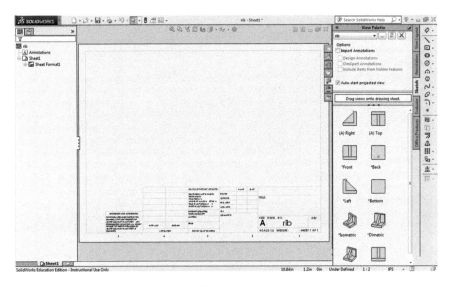

Fig. 12.8 (with permission from Dassault Systems)

11. Drag views onto Drawing Sheet is highlighted in yellow. Click on the front view and drag it to drawing sheet.
12. Set the properties for the new view or select a location to place it.

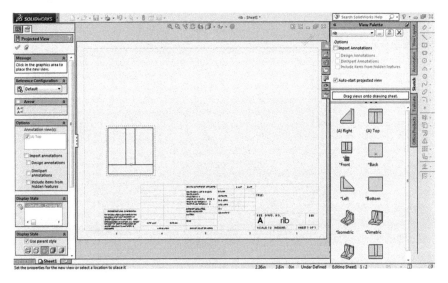

Fig. 12.9 (with permission from Dassault Systems)

13. Now drag the cursor to the left side, left view will automatically seen on the drawing sheet.

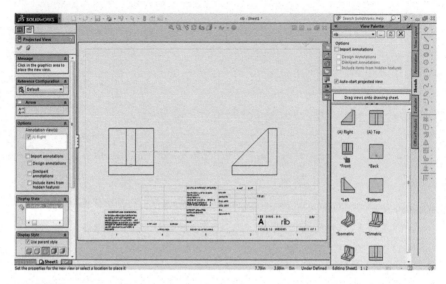

Fig. 12.10 (with permission from Dassault Systems)

14. In the same way, drag the cursor on the top to see the top view and side view as can be seen in the figure given below.

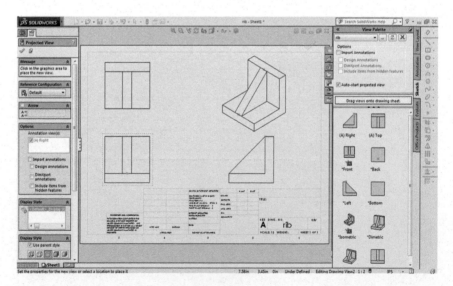

Fig. 12.11 (with permission from Dassault Systems)

15. Click ok.
16. Now from the feature manager design tree select **Sheet 1** and right-click to see the **properties**. After clicking on the properties, it can be seen that the drawing was created in **three angle projection**. Edit it to **first angle projection**. Click **ok**.

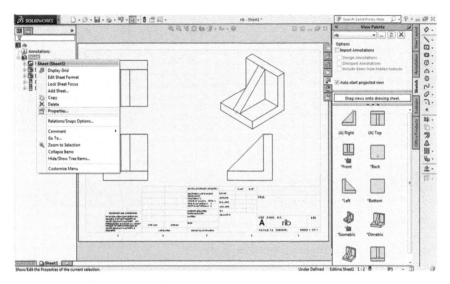

Fig. 12.12 (with permission from Dassault Systems)

17. In the Annotations toolbar, select **Smart Dimension** to create dimension for one or more selected entities.

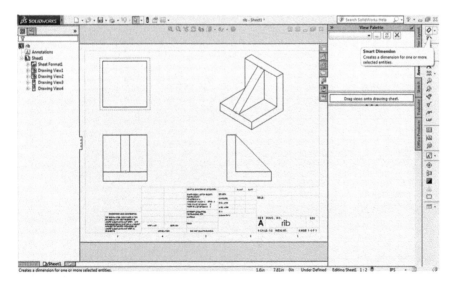

Fig. 12.13 (with permission from Dassault Systems)

18. Give the dimension to the drawing and click **ok** after completion.

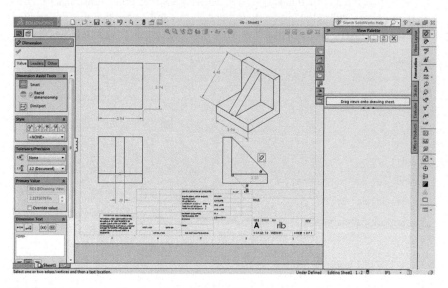

Fig. 12.14 (with permission from Dassault Systems)

19. In the **Annotation** toolbar, **Model Items** icon is available. Click it.
20. In the **source/destination**, select the entire model by editing the source.
21. In **Dimensions** section, click the icon stating Not Marked for Drawing.

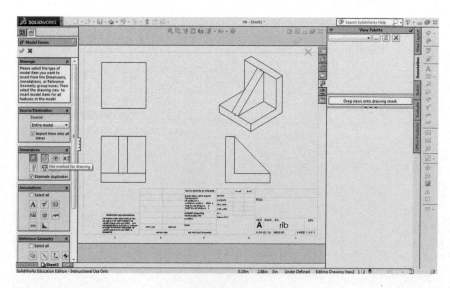

Fig. 12.15 (with permission from Dassault Systems)

22. Click **ok**. Model items select the entire model and give dimensions to each part or view which can be seen onto drawing sheet.

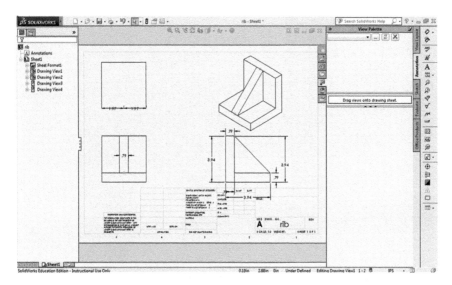

Fig. 12.16 (with permission from Dassault Systems)

23. Minimize the size by zoom in and save the drawing sheet by clicking on **File > Save as**.
24. Name the file and click **Save**.

Chapter 13
To Create a Flanged Coupling

The present chapter is about creation of an assembly drawing of an engineering component from scratch. It is basically the summation of all the previous chapters.

1. Enter the **Part** module of the SOLIDWORKS. **Right-click** the **Front Plane** and select **Sketch**. Draw the hexagon with the inscribed circle of the desired dimensions using the **Polygon** option in the sketch window.

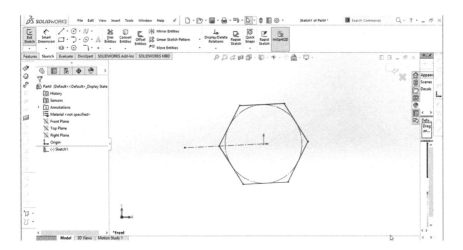

Fig. 13.1 (with permission from Dassault Systems)

2. Select **Smart Dimensions** and enter the **diameter** of the circle as **22 mm** under the **Primary Value** field of the **Dimensions**.

© Springer Nature Switzerland AG 2020   261
K. Kumar et al., *Mastering SolidWorks*, Management and Industrial Engineering,
https://doi.org/10.1007/978-3-030-38901-7_13

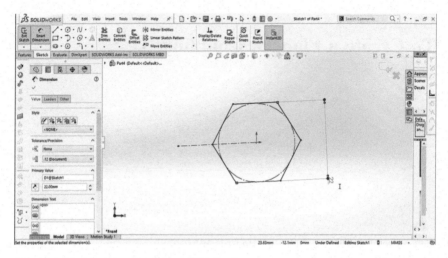

Fig. 13.2 (with permission from Dassault Systems)

3. Click on **Feature** and then select **Extruded Boss/Base** to enter the depth of the hexagon created. Select **Mid Plane** in **Direction 1** field. Enter the **depth** as **10 mm** under the **Direction 1** Field.

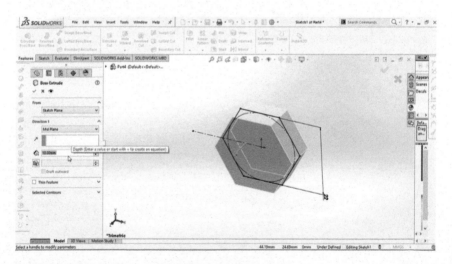

Fig. 13.3 (with permission from Dassault Systems)

4. Click on ✓ option to **terminate** the **Extruded Boss/Base** option. Select the **base** of the created part as the **sketching plane**.

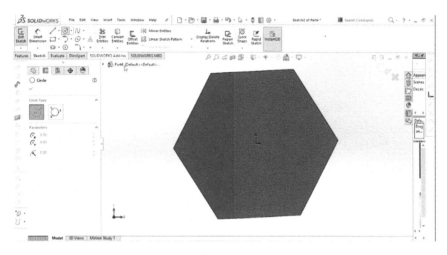

Fig. 13.4 (with permission from Dassault Systems)

5. Next, the **hole** will be created. Select the **frontal plane** of the part. Click on **Sketch** and then select **Circle**. Select **Two Point Circle** under **Circle Type**. Select the center of the circle by clicking the mouse, drag the mouse and terminate the process by another mouse click. Enter the diameter of the circle as 11 mm in the **Parameters** field. Click on ✓ to exit the sketch mode.

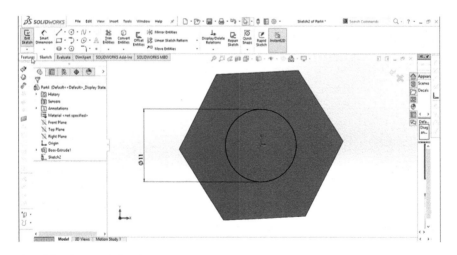

Fig. 13.5 (with permission from Dassault Systems)

6. Click on **Features** and then select **Extruded Cut** option. Select **Through All** option under **Direction 1 field**. Click on ✓ to create the hole of the desired dimension.

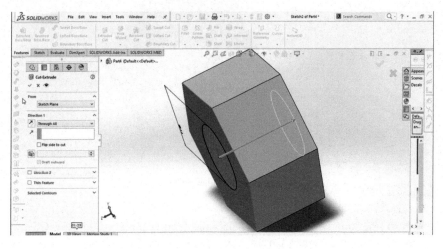

Fig. 13.6 (with permission from Dassault Systems)

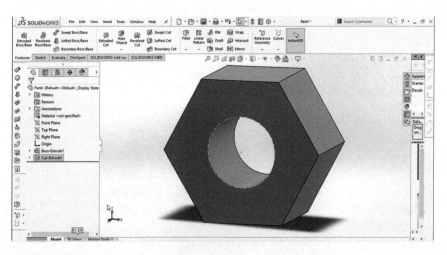

Fig. 13.7 (with permission from Dassault Systems)

7. Select the **frontal plane** of the part. Click on **Sketch** and select **Two Point Circle** under the **Circle Type** of the **Circle Window** by specifying the **center of the circle** and end point as one of the **edges of the hexagon**.

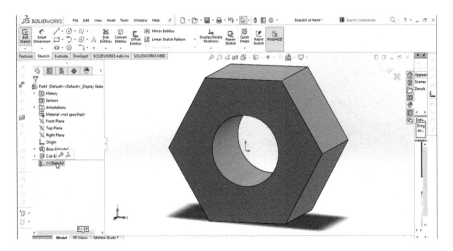

Fig. 13.8 (with permission from Dassault Systems)

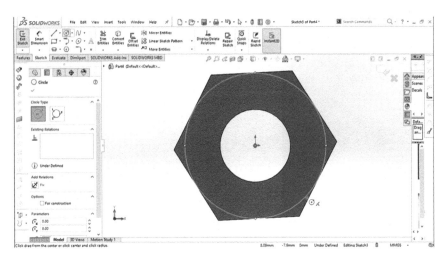

Fig. 13.9 (with permission from Dassault Systems)

8. Select Features and then click on **Extruded Cut**. Select **Blind** and if necessary click on the **Flip sides to out** to flip the direction of cut. Specify the depth of the extrude cut as desired. Enter Draft as 60° under Direction 1 field. Click on ✓ to terminate the **Extruded Cut** option.

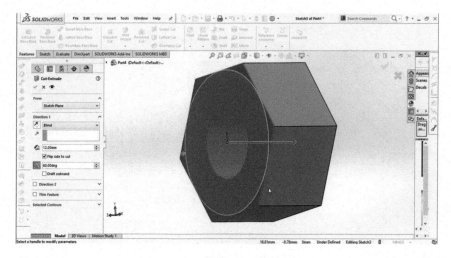

Fig. 13.10 (with permission from Dassault Systems)

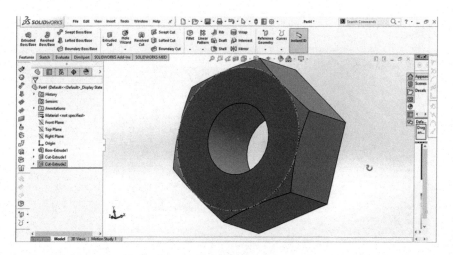

Fig. 13.11 (with permission from Dassault Systems)

9. Click on **File**. Select **Save As…** option. Enter the file name as **nut** in the **File name** field. Click on **Save** to save the created part.

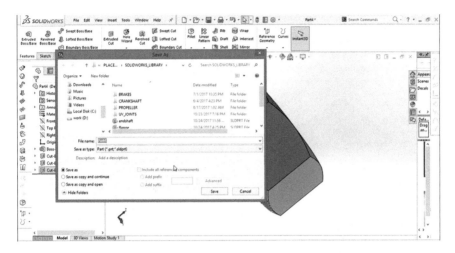

Fig. 13.12 (with permission from Dassault Systems)

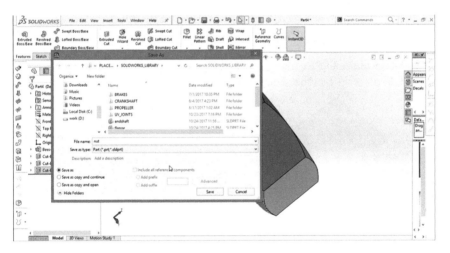

Fig. 13.13 (with permission from Dassault Systems)

10. Next, a bolt will be created. Click on **File** and select **New** to create a new part in
    the **Part module** of the SOLIDWORKS. Right-click on **Front Plane** and select
    Sketch.

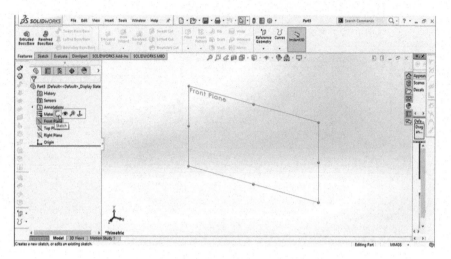

Fig. 13.14 (with permission from Dassault Systems)

11. Click on **Sketch** and then select **Circle**. Select **Two Point Circle** under **Circle Type**. Select the center of the circle by clicking the mouse, drag the mouse and terminate the process by another mouse click. Enter the diameter of the circle as 11 mm under the **Primary Value** field using the **Smart Dimensions** option. Click on ✓ to exit the sketch mode.

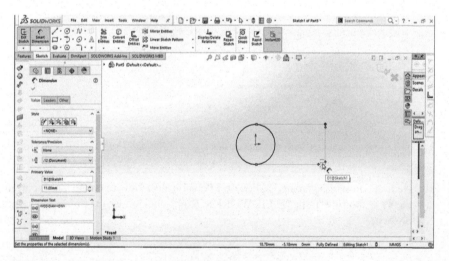

Fig. 13.15 (with permission from Dassault Systems)

12. Click on **Features** and then select **Extruded Boss/Base**. Select **Mid Plane** option under **Direction 1** window. Enter the depth as 46 mm. Click on ✓ to terminate the Extruded Boss/Base option.

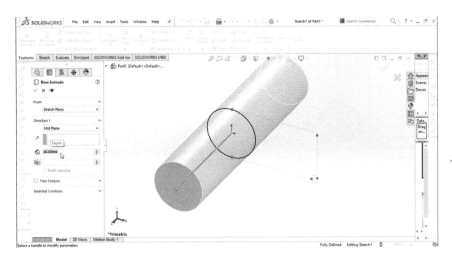

Fig. 13.16 (with permission from Dassault Systems)

13. Select the one end of the created shaft as the sketching plane.

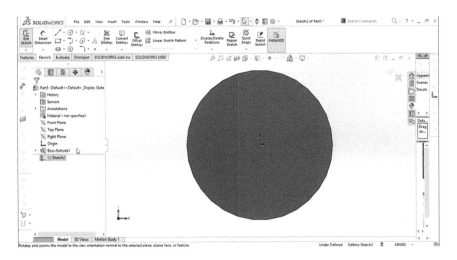

Fig. 13.17 (with permission from Dassault Systems)

14. Click on **Polygon** and create a hexagon with the inscribed circle of diameter 22 mm. Click on ✓ to terminate the process.

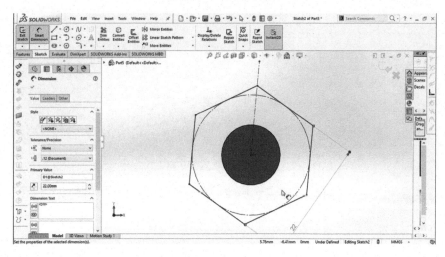

Fig. 13.18 (with permission from Dassault Systems)

15. Click on **Feature** and select on **Extruded Boss/Base** option. Select **Blind** option in the **Direction 1** window. Enter the depth of extrude as 12 mm. Click on ✓ to terminate the process.

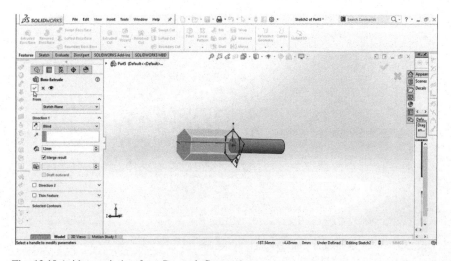

Fig. 13.19 (with permission from Dassault Systems)

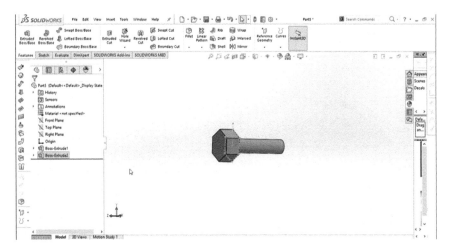

Fig. 13.20 (with permission from Dassault Systems)

16. Select the **frontal plane** of the created hexagon as the sketching plane.

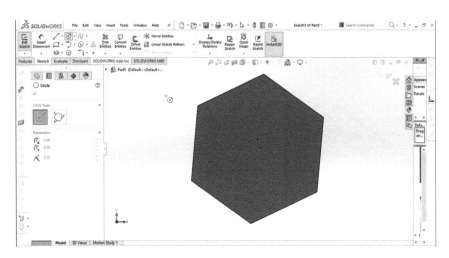

Fig. 13.21 (with permission from Dassault Systems)

17. Click on **Sketch** and then select **Circle**. Select **Two Point Circle** under **Circle Type**. Select the center of the circle by clicking the mouse, drag the mouse and terminate the process by another mouse click. Enter the diameter of the circle as 22 mm under the **Parameters** field. Click on ✓ to exit the sketch mode.

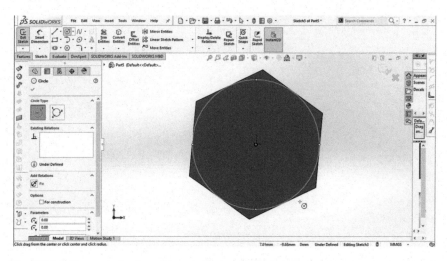

Fig. 13.22 (with permission from Dassault Systems)

18. Select Features and then click on **Extruded Cut**. Select **Blind** and if necessary click on the **Flip sides to out** to flip the direction of cut. Specify the depth of the extrude cut as desired. Enter Draft as 60° under Direction 1 field. Click on ✓ to terminate the **Extruded Cut** option.

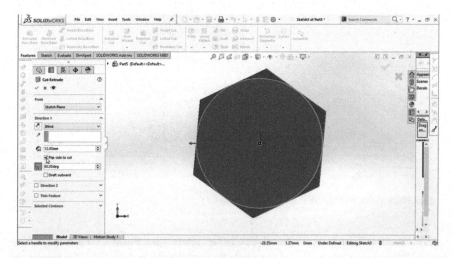

Fig. 13.23 (with permission from Dassault Systems)

19. Save the created part as **Bolt**.

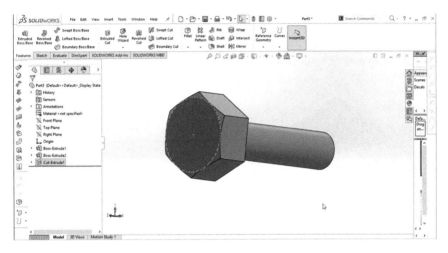

Fig. 13.24 (with permission from Dassault Systems)

20. Next, a taper pin will be created. Select the **Front Plane** for the sketch. Click on
    Sketch and select **Rectangle**. Create a rectangle by specifying the intersection
    point of its diagonal as the first point and a point on one of its corner as the
    other point (**Second rectangle** under **Rectangle Type**).

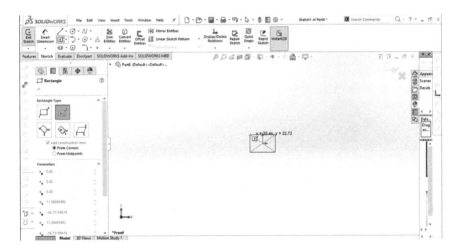

Fig. 13.25 (with permission from Dassault Systems)

21. Use the **Smart Dimensions** option to specify the dimensions of the created rectangle. Click on ✔ to terminate the process.

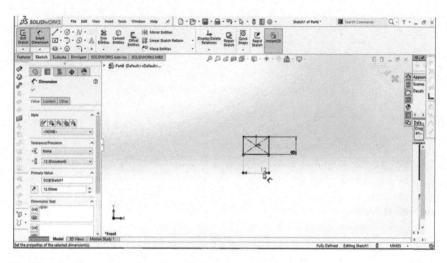

Fig. 13.26 (with permission from Dassault Systems)

22. Click on **Features** and select **Extruded Boss/Base option**. Select the **Mid Plane** option from the **Direction 1**. Enter the depth of extrude as 55 mm. Click on ✔ to terminate the process.

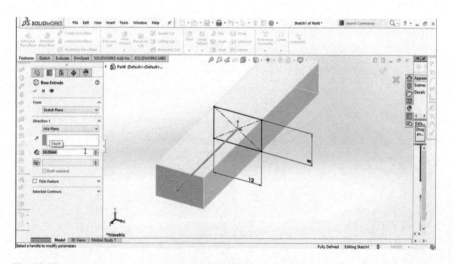

Fig. 13.27 (with permission from Dassault Systems)

23. The taper pin has been created. **Save** the created part as **Taper Pin**.

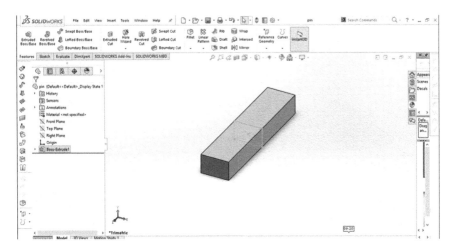

Fig. 13.28 (with permission from Dassault Systems)

24. Next, an end shaft will be created. Click on **Sketch** and then select **Circle**. Select **Two Point Circle** under **Circle Type**. Select the center of the circle by clicking the mouse, drag the mouse and terminate the process by another mouse click. Enter the diameter of the circle as 30 mm under the **Primary Value** field using the **Smart Dimensions** option. Click on ✓ to exit the sketch mode.
25. Click on **Features** and then select **Extruded Boss/Base**. Select **Mid Plane** option under **Direction 1** window. Enter the depth as 50 mm. Click on ✓ to terminate the Extruded Boss/Base option.

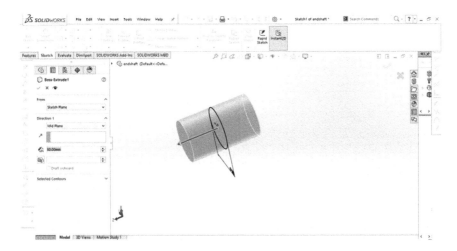

Fig. 13.29 (with permission from Dassault Systems)

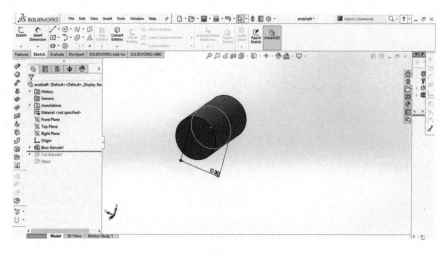

Fig. 13.30 (with permission from Dassault Systems)

26. Select **one end face** of the created part as the **sketching plane**. Create a sketch on this plane as shown in the figure below.

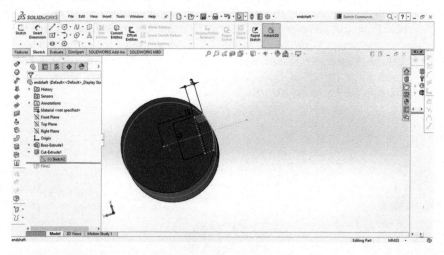

Fig. 13.31 (with permission from Dassault Systems)

27. Select **Features** and click on **Extruded Cut**. Select the sketch created above for the extruded cut. Select **Blind** under **Direction 1** and enter the **depth of cut as 40 mm**.

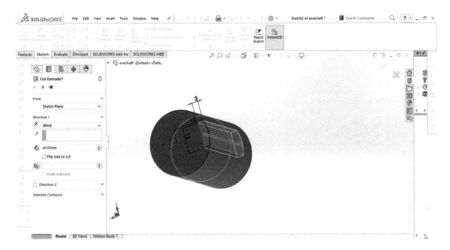

Fig. 13.32 (with permission from Dassault Systems)

28. Click on **Features** and select the **Fillet** option. Select the **highlighted edge** in the **Items To Fillet** field. Enter the **fillet radius as 5 mm** under the **Fillet Parameters** field. Click on ✓ to terminate the process. Save the part created as **End Shaft**.

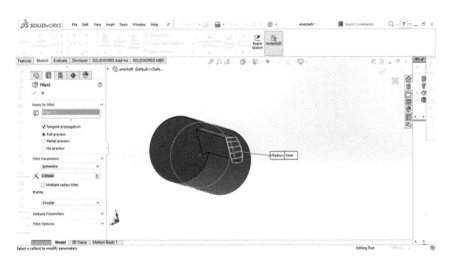

Fig. 13.33 (with permission from Dassault Systems)

29. Next, the **Assembly** will be done to generate a Flange Coupling. Click on **File** and select **New**. Then click on **Assembly** module in the **New SOLIDWORKS Document** window. Click on **OK**.

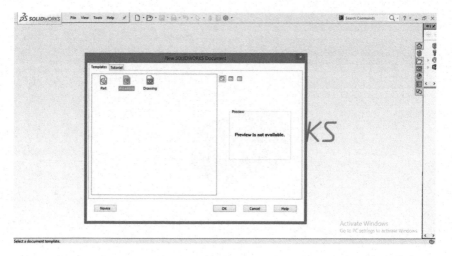

Fig. 13.34 (with permission from Dassault Systems)

30. Select **Assembly** and Click on **Insert Components** and select the **Flanges** created.

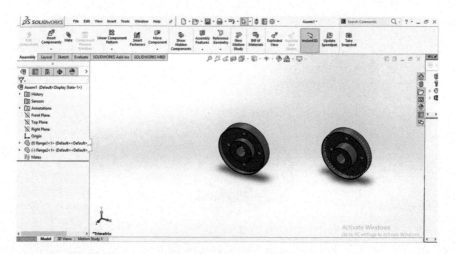

Fig. 13.35 (with permission from Dassault Systems)

31. Click on **Assembly** and select **Move Component** to rotate or translate the flanges as required. Select **Mates** and then select the edges of the flanges. The selected edges will be shown in the **Mate Selection** field. Next, click on **Coincident** under **Standard Mates**.

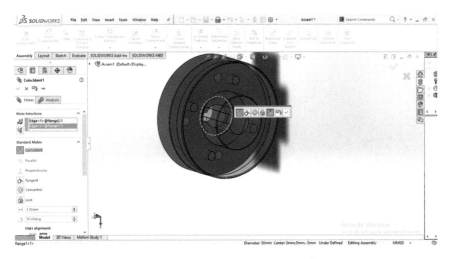

Fig. 13.36 (with permission from Dassault Systems)

32. Next, select the **faces of the two flanges**. Click on **Coincident** under the **Standard Mates** option. Click on ✓ to terminate the process. The two flanges will be assembled as shown.

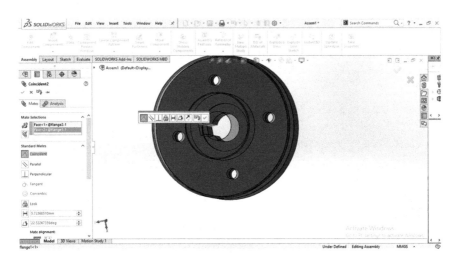

Fig. 13.37 (with permission from Dassault Systems)

33. Next, click on **Insert Components**. Select the **nut** and the **bolt** from the list.

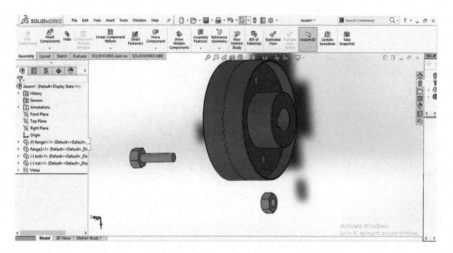

Fig. 13.38 (with permission from Dassault Systems)

34. Click on **Move Components** and select **Rotate Component**. Rotate the bolt if
    necessary. Next, Select **Mate**. Select the **shaft end of the bolt** as one of the
    faces and the **inner face of the hole** in the flange as the second face. The
    selections made will be shown in the Mate Selections field. Select **Centric**
    under the **Standard Mates**.
35. Next, select the **face of the bolt head** (the face which joins with the cylindrical
    end of the bolt) and then select the **face of the flange**. Select **Coincident** under
    Standard Mates. The bolt will then move to take its place as shown in the
    figure.

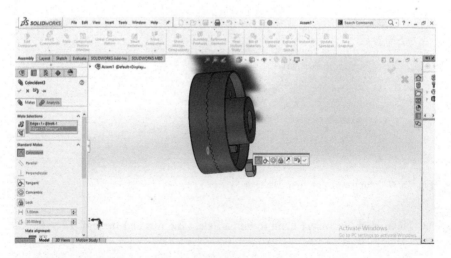

Fig. 13.39 (with permission from Dassault Systems)

36. Next, click on **Move Component** and then click on **Rotate** to rotate the nut if required.

37. Click on **Mate** and select the **inner face of the hole in the nut** and the **cylindrical part of the bolt** as the second face. Click on **Concentric** under **Standard Mates**.

38. Select **face of the nut as one of the faces** and then the **face of the flange as the second face**. Click on **Coincident** to complete the assembly of nut with the bolt.

Fig. 13.40 (with permission from Dassault Systems)

39. For the assembly of other bolt and nut assembly, **Circular Pattern** option may be used. Select **Linear Component Pattern** and then click on **Circular Pattern**. Select the **edge of the hole** in the flange as one of the **Parameters**. Enter the **angle for pattern** as **90°**.

40. Select the **bolt and nut assembly** already created under **Components to Pattern**. Click on ✓ to terminate the process.

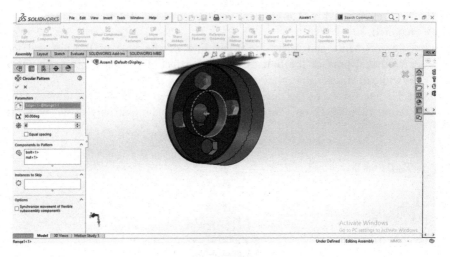

Fig. 13.41 (with permission from Dassault Systems)

41. Click on **Insert Component**. Select the **End Shaft** as the other component to be assembled.

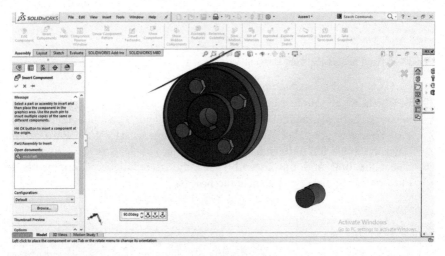

Fig. 13.42 (with permission from Dassault Systems)

42. Select **Mate**. Next, select the **periphery of the end shaft** as **one of the edges** and the **periphery of the flange hole** in which the end shaft is to be inserted as the **second edge**. Select **Coincident** under the **Mate Selection**.

43. Next, select the **cylindrical face of the end shaft** and then the **inner face of the flange hole**. The two selected faces will be reflected in the **Mate Selections** window. Click on **Coincident** to generate the required assembly.

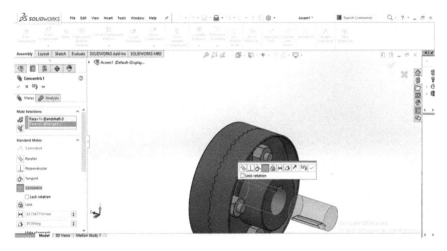

Fig. 13.43 (with permission from Dassault Systems)

44. The assembly created for the end shaft with the flange is shown in the figure. Repeat steps 41–43 for the assembly of end shaft with other flange.

Fig. 13.44 (with permission from Dassault Systems)

45. Next, the **taper pin** will be **assembled**. Click on **Insert Component** and select the **pin** part from the directory.

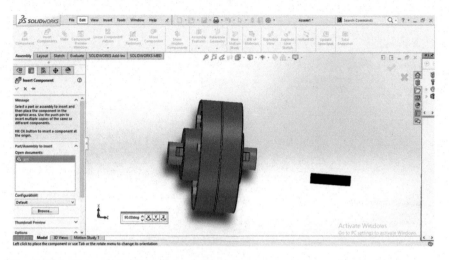

Fig. 13.45 (with permission from Dassault Systems)

46. Click on **Mate** which will open the **Mate Selections** window. Select the **lower face of the pin** as one of the faces and then select the **upper flat face of the end shaft** as depicted in the figure.
47. Click on **Coincident** under **Standard Mates** to align the pin with the flange as shown in the figure.

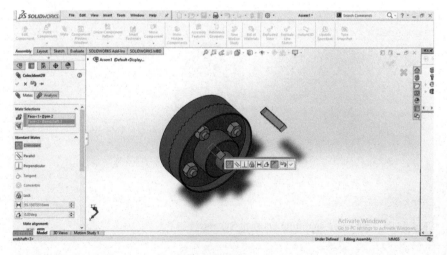

Fig. 13.46 (with permission from Dassault Systems)

48. Next, select the **face of the flange** as one of the faces and the **end face of the pin** as the second face. Click on **Coincident** to assemble the pin with the flange.

Fig. 13.47 (with permission from Dassault Systems)

49. Select the **edges of the pin** and the **end shaft** as shown in the figure. Click on **Coincident** to complete the assembly.

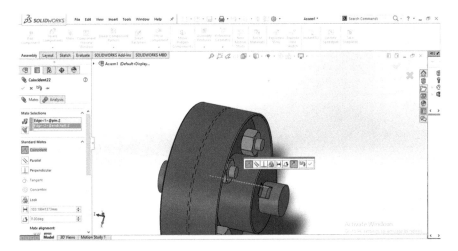

Fig. 13.48 (with permission from Dassault Systems)

50. Click on **Exploded View** option to get the expanded view of the assembly.

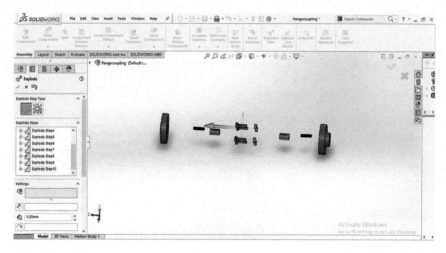

Fig. 13.49 (with permission from Dassault Systems)

Chapter 14
To Create a Foot Step Bearing

The present chapter is similar to earlier chapter and the steps are also similar. The various steps are provided in the figures below.

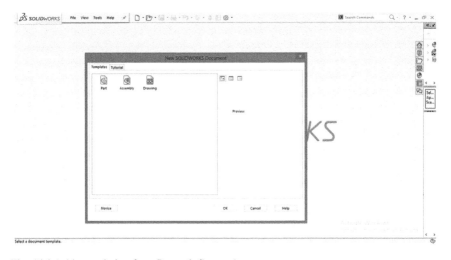

Fig. 14.1 (with permission from Dassault Systems)

© Springer Nature Switzerland AG 2020
K. Kumar et al., *Mastering SolidWorks*, Management and Industrial Engineering,
https://doi.org/10.1007/978-3-030-38901-7_14

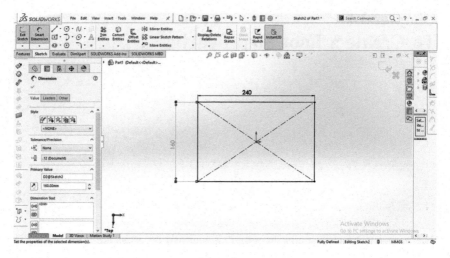

Fig. 14.2 (with permission from Dassault Systems)

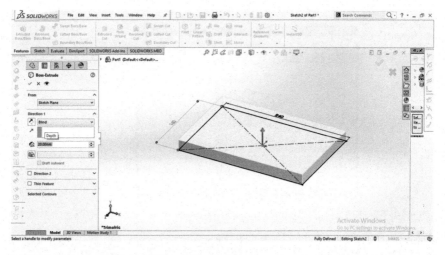

Fig. 14.3 (with permission from Dassault Systems)

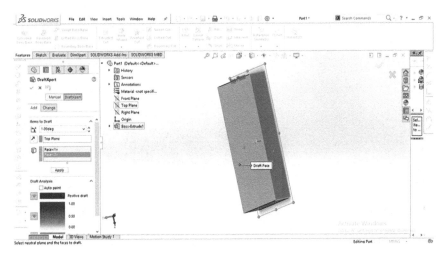

Fig. 14.4 (with permission from Dassault Systems)

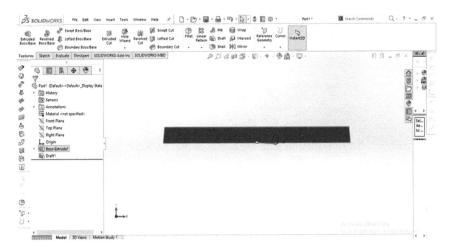

Fig. 14.5 (with permission from Dassault Systems)

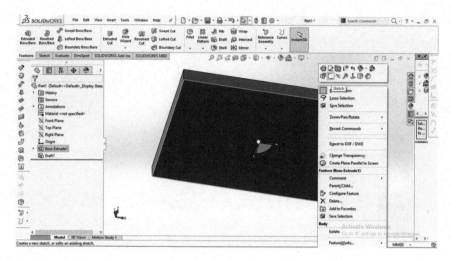

Fig. 14.6 (with permission from Dassault Systems)

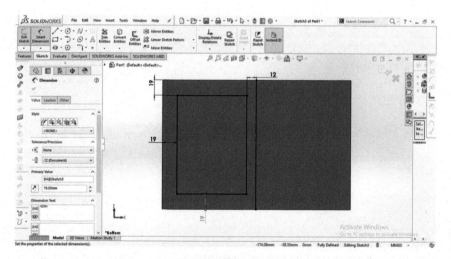

Fig. 14.7 (with permission from Dassault Systems)

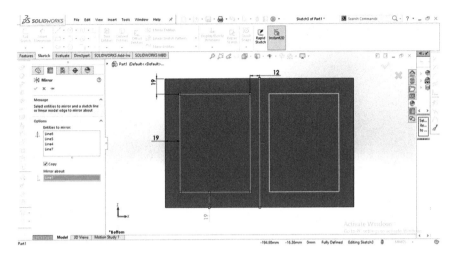

Fig. 14.8 (with permission from Dassault Systems)

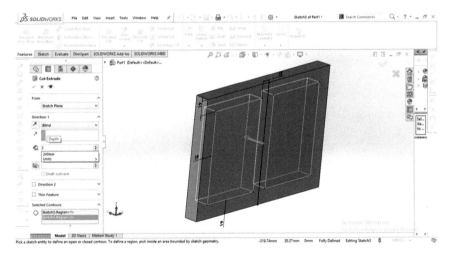

Fig. 14.9 (with permission from Dassault Systems)

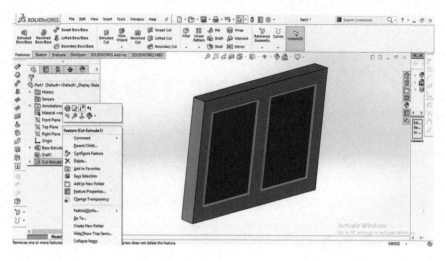

Fig. 14.10 (with permission from Dassault Systems)

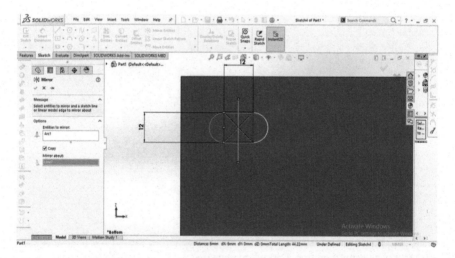

Fig. 14.11 (with permission from Dassault Systems)

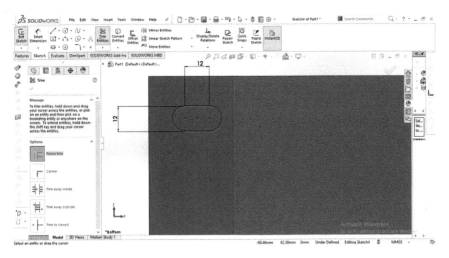

Fig. 14.12 (with permission from Dassault Systems)

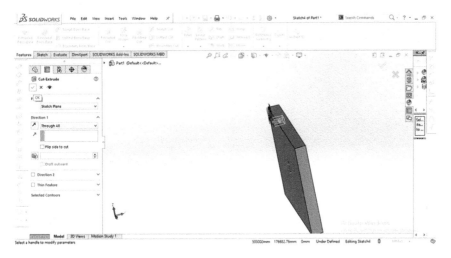

Fig. 14.13 (with permission from Dassault Systems)

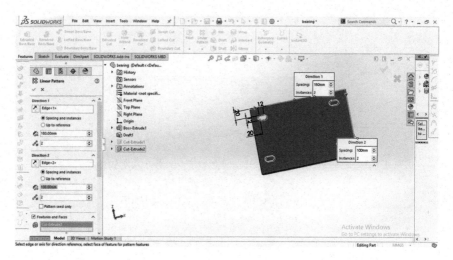

Fig. 14.14 (with permission from Dassault Systems)

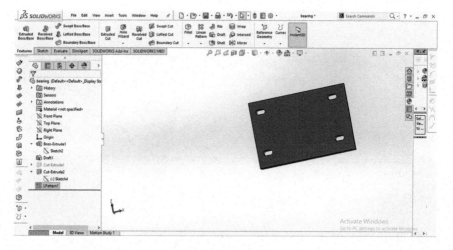

Fig. 14.15 (with permission from Dassault Systems)

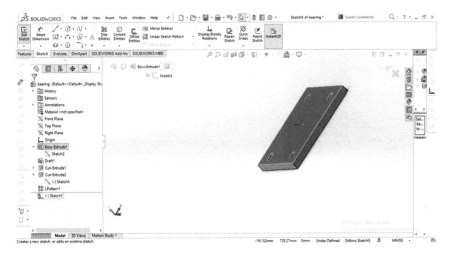

Fig. 14.16 (with permission from Dassault Systems)

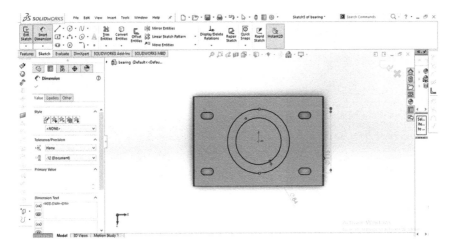

Fig. 14.17 (with permission from Dassault Systems)

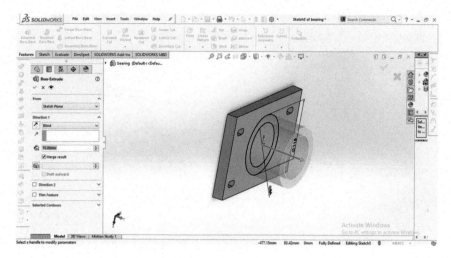

Fig. 14.18 (with permission from Dassault Systems)

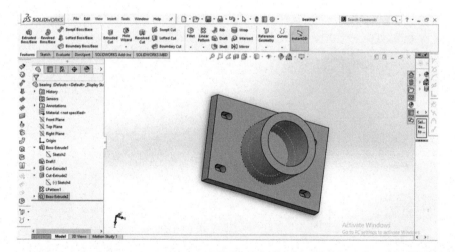

Fig. 14.19 (with permission from Dassault Systems)

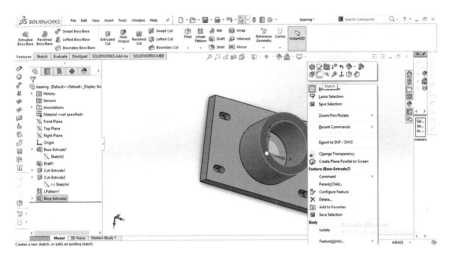

Fig. 14.20 (with permission from Dassault Systems)

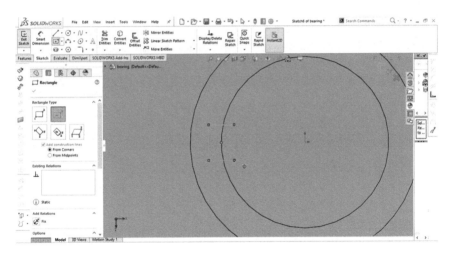

Fig. 14.21 (with permission from Dassault Systems)

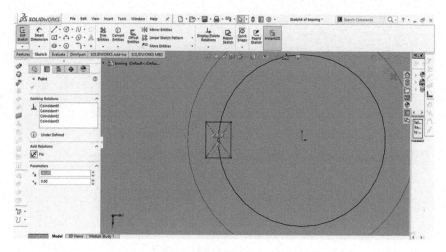

Fig. 14.22 (with permission from Dassault Systems)

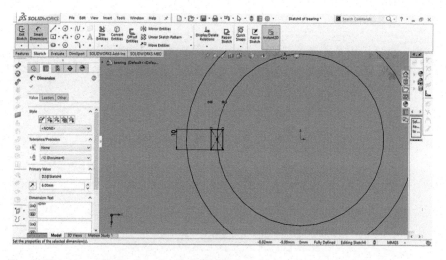

Fig. 14.23 (with permission from Dassault Systems)

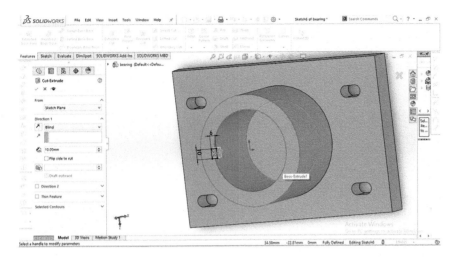

Fig. 14.24 (with permission from Dassault Systems)

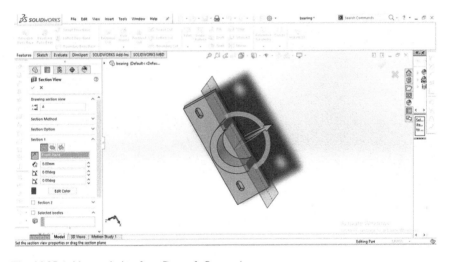

Fig. 14.25 (with permission from Dassault Systems)

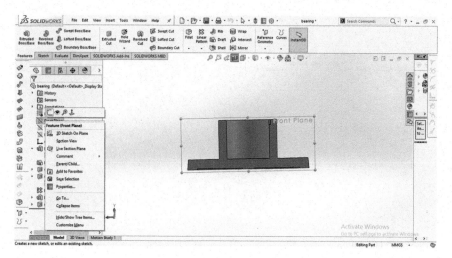

Fig. 14.26 (with permission from Dassault Systems)

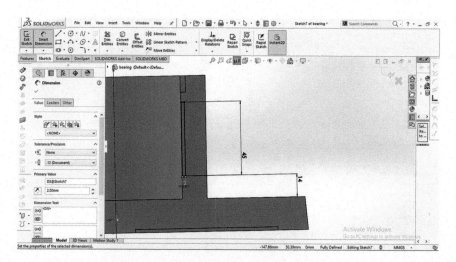

Fig. 14.27 (with permission from Dassault Systems)

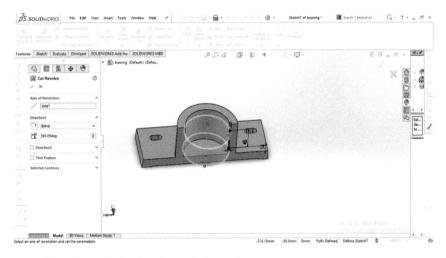

Fig. 14.28 (with permission from Dassault Systems)

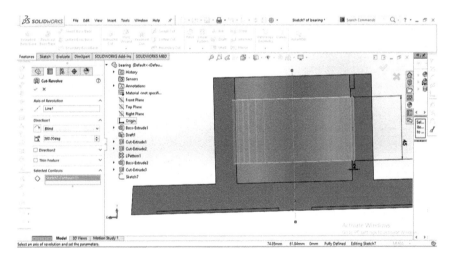

Fig. 14.29 (with permission from Dassault Systems)

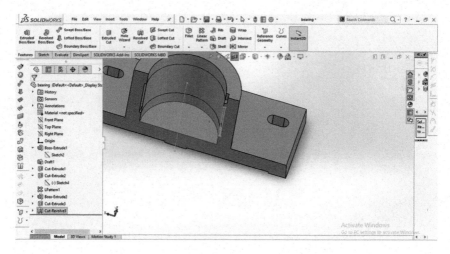

Fig. 14.30 (with permission from Dassault Systems)

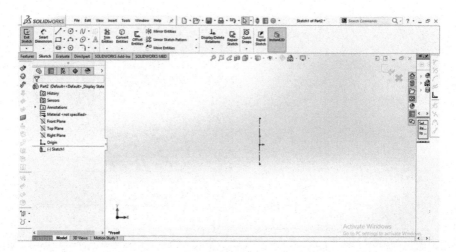

Fig. 14.31 (with permission from Dassault Systems)

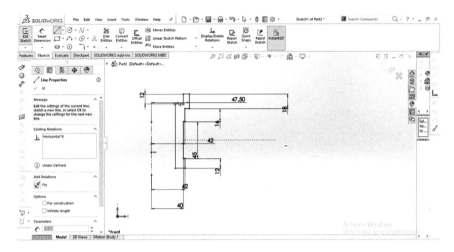

Fig. 14.32 (with permission from Dassault Systems)

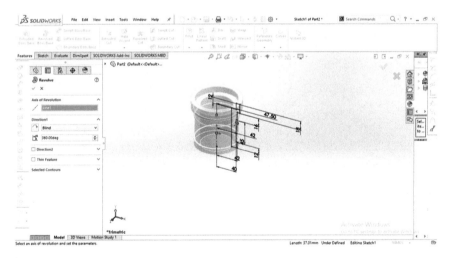

Fig. 14.33 (with permission from Dassault Systems)

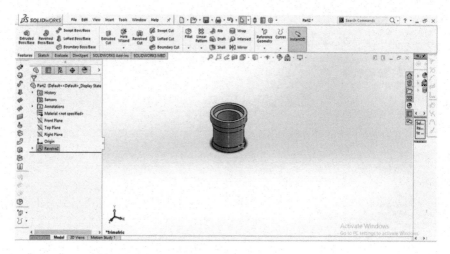

Fig. 14.34 (with permission from Dassault Systems)

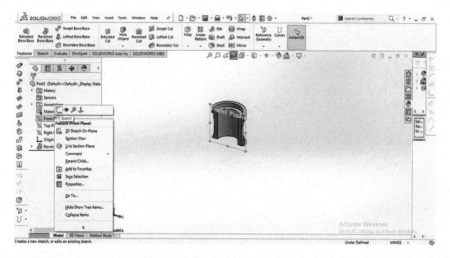

Fig. 14.35 (with permission from Dassault Systems)

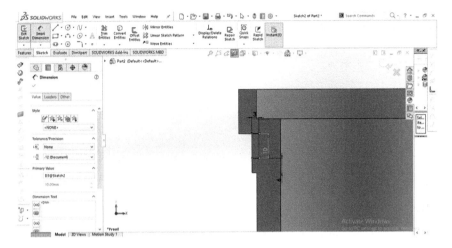

Fig. 14.36 (with permission from Dassault Systems)

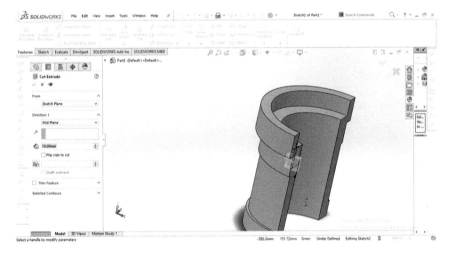

Fig. 14.37 (with permission from Dassault Systems)

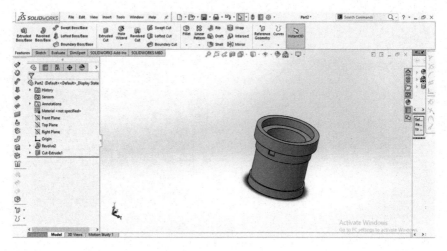

Fig. 14.38 (with permission from Dassault Systems)

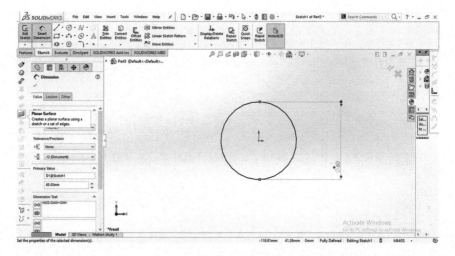

Fig. 14.39 (with permission from Dassault Systems)

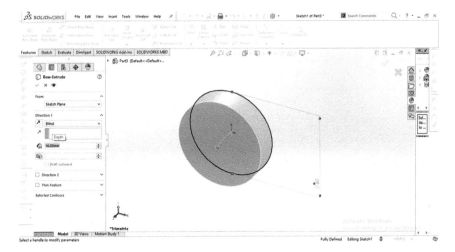

Fig. 14.40 (with permission from Dassault Systems)

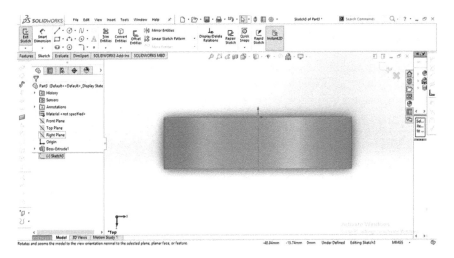

Fig. 14.41 (with permission from Dassault Systems)

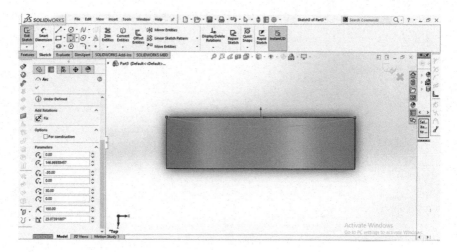

Fig. 14.42 (with permission from Dassault Systems)

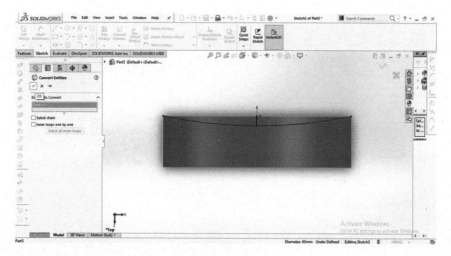

Fig. 14.43 (with permission from Dassault Systems)

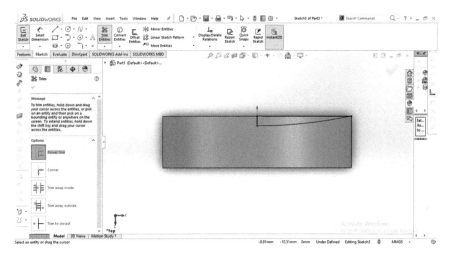

Fig. 14.44 (with permission from Dassault Systems)

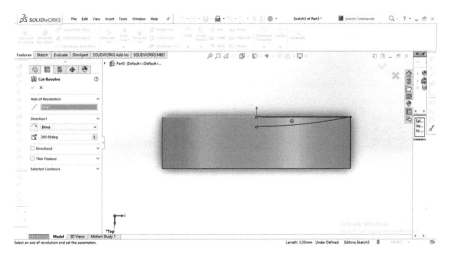

Fig. 14.45 (with permission from Dassault Systems)

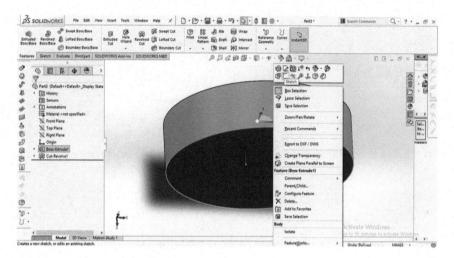

Fig. 14.46 (with permission from Dassault Systems)

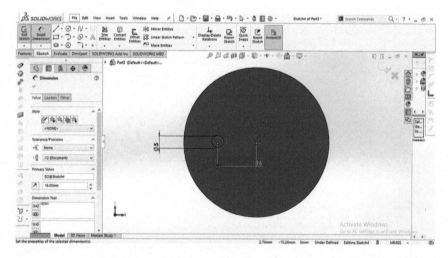

Fig. 14.47 (with permission from Dassault Systems)

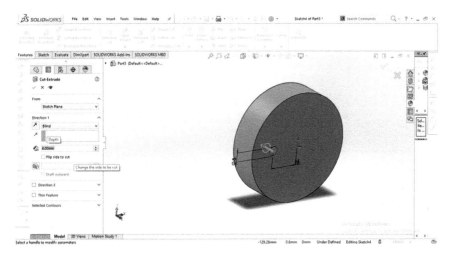

Fig. 14.48 (with permission from Dassault Systems)

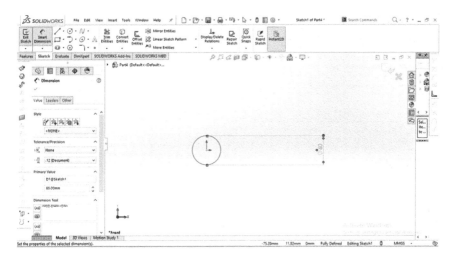

Fig. 14.49 (with permission from Dassault Systems)

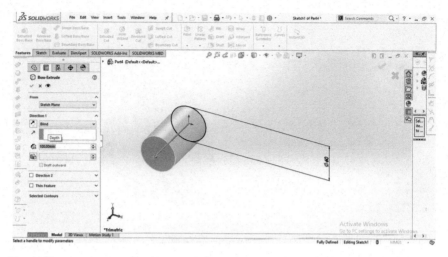

Fig. 14.50 (with permission from Dassault Systems)

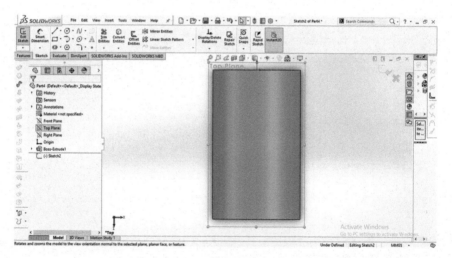

Fig. 14.51 (with permission from Dassault Systems)

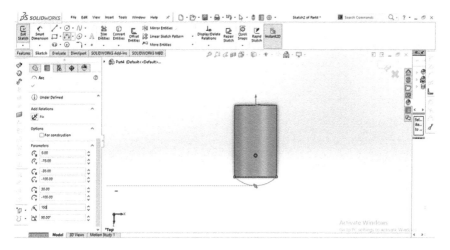

Fig. 14.52 (with permission from Dassault Systems)

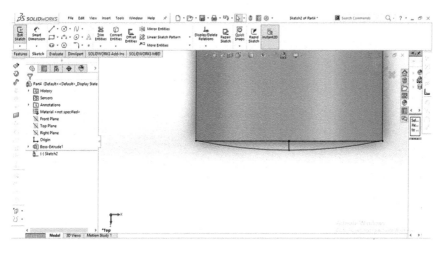

Fig. 14.53 (with permission from Dassault Systems)

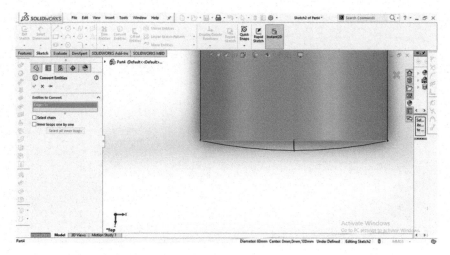

Fig. 14.54 (with permission from Dassault Systems)

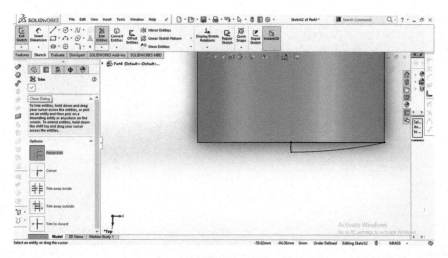

Fig. 14.55 (with permission from Dassault Systems)

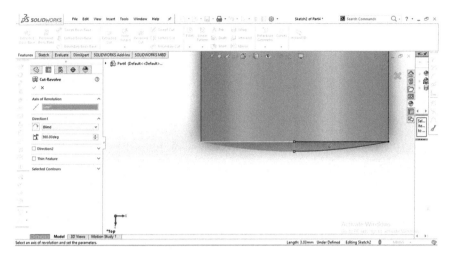

Fig. 14.56 (with permission from Dassault Systems)

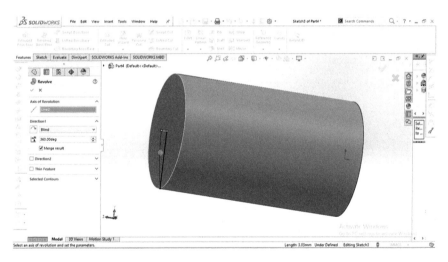

Fig. 14.57 (with permission from Dassault Systems)

Fig. 14.58 (with permission
from Dassault Systems)

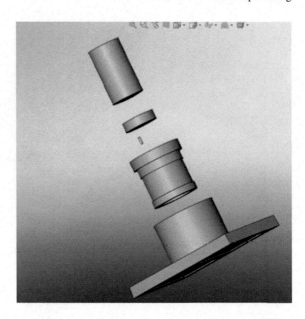

Fig. 14.59 (with permission
from Dassault Systems)

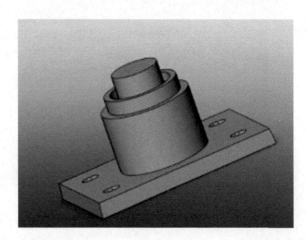

Printed by Printforce, the Netherlands